Data
Grab

The Costs of Connection:
How Data Is Colonizing Human Life and
Appropriating It for Capitalism

Data Grab

The New Colonialism of Big Tech and How to Fight Back

Ulises A. Mejias
& Nick Couldry

The University of Chicago Press

The University of Chicago Press, Chicago 60637

Published 2024

Printed in the United States of America

33 32 31 30 29 28 27 26 25 24   1 2 3 4 5

ISBN-13: 978-0-226-83230-2 (cloth)
ISBN-13: 978-0-226-83231-9 (e-book)
DOI: https://doi.org/10.7208/chicago/9780226832319.001.0001

First published in the United Kingdom by WH Allen, an imprint of Ebury Publishing/Penguin Random House UK, 2024.

Library of Congress Cataloging-in-Publication Data

Names: Mejias, Ulises Ali, author. | Couldry, Nick, author.
Title: Data grab : the new colonialism of big tech and how to fight back / Ulises A. Mejias and Nick Couldry.
Description: Chicago, IL : The University of Chicago Press, 2024. | Includes bibliographical references and index.
Identifiers: LCCN 2023039003 | ISBN 9780226832302 (cloth) | ISBN 9780226832319 (ebook)
Subjects: LCSH: Internet industry—Social aspects. | Data privacy. | Data sovereignty. | Corporate power. | Information technology—Social aspects. | Electronic data processing—Social aspects.
Classification: LCC HD9696.8.A2 M455 2024 | DDC 338.4/7004678—dc23/eng/20230929
LC record available at https://lccn.loc.gov/2023039003

♾ This paper meets the requirements of ANSI/NISO Z39.48-1992 (Permanence of Paper).

To the members of Tierra Común

CONTENTS

ACKNOWLEDGEMENTS

THIS BOOK WAS written in an unusually intense period between June 2022 and mid-June 2023. It draws conceptually on the argument of our previous book, *The Costs of Connection: How Data is Colonizing Human Life and Appropriating it for Capitalism* (2019, Stanford University Press). The story is completely retold here, because our thinking has inevitably evolved in response to the authors of various published responses and the members of nearly a hundred live and virtual audiences to whom we have spoken since late 2018 in Africa, Asia, Europe, the Americas and Australasia. We are grateful to all of them.

We are also thankful to Nabil Echchaibi and Paola Ricaurte, who kindly read individual chapters, and above all to Isobel Edwards and Louise Edwards who were willing to read the whole manuscript in its first version and provide very useful comments. Thanks also to Miriyam Aouragh, Benedetta Brevini, Lilly Irani, Ilona Kickbusch, Sebastian Lehuedé and Ella McPherson, who advised on specific points during the writing or build-up to the book. Huge thanks in particular to Nick's research assistant, Louise Marie Hurel at the London School of Economics and Political Science, for her skilful and prompt research – we couldn't have done this without her.

We have also benefitted enormously from the belief, encouragement and feedback of Jamie Joseph, our editor at WH Allen, and from the expert comments on various versions of the manuscript by Amanda Waters at WH Allen and Joe Calamia at Chicago University Press, our US publisher and long-term supporter. Thanks also to Ross Jamieson, Ian Allen and Ben Murphy for help in copy editing, proofreading and indexing.

On a personal note, Nick would like to thank the students on his Media, Data and Social Order course at LSE for their inspiration and thoughtful challenges, and to express his deep gratitude to Louise Edwards for her love, support and belief during what has been an extremely arduous and stressful period for both of us. Ulises would like to thank friends and family who provided instrumental support and welcome distraction during the final intense stages of writing: Lisa Dundon, Zillah Eisenstein, Jenny Rosenberg, Winfried and Sybille Thaa, and Demir and Lale Barlas. Most importantly, he would like to thank Asma Barlas for providing a loving, nurturing and critical relationship in which his ideas have developed – truly, a home for his heart, mind and soul.

Finally, we want to mention the international community of Tierra Común that, with our dear friend Paola Ricaurte, we have had the privilege of founding and seeing grow since June 2020. It has been a great joy to see this network of thinkers and activists gain momentum in the fight to resist data colonialism, and our meetings virtual and now, fortunately, in person, have been a continual inspiration. We therefore dedicate this book to the members of Tierra Común.

A note about author order: Our fruitful collaboration has been, from the start, the result of joint work to which we have equally contributed. In order to better represent this, we prefer to rotate the order in which our names are listed, rather than going each time with the conventional alphabetical order.

Ulises Ali Mejías (Ithaca, US)   Nick Couldry (Islip, UK)
June 2023

FROM LANDGRAB
TO DATA GRAB

THE ADVISORS TO King Lobengula were suspicious of the telegraph wires being stretched across their land by the British South Africa Company in the late nineteenth century. They believed the white men's plan was to use the wires to tie and restrain their king, ruler of the Northern Ndebele people in Matabeleland. Even when the official purpose of the telegraph was explained to them, they were still dismissive. Why would such a thing be needed, they asked, when they already possessed effective means of long-distance communication such as drums and smoke signals?

To many, this might sound like a familiar story: the story of pre-modern people standing in the way of inevitable progress, or the story of misguided resistance to a technology that eventually paved the way to a better world.

But the advisors to King Lobengula were justified in their suspicions. The British South Africa Company, led by Cecil Rhodes, declared war against the Ndebele in 1893, and continued with the suppression of the Matabeleland and Mashonaland uprisings in 1896. One of the pretexts used to wage war was that the locals were stealing the copper wire to make ornaments and hunting tools. The telegraph was important for other reasons too. From a military perspective, it would prove to be crucial for orchestrating the colonisation of southern Africa, including what would become southern Rhodesia (now Zimbabwe). It would have been much more difficult to coordinate troop movements and send alerts without it. As a result of those wars, by 1930 about 50 per cent of the country's land – 49 million acres – had been granted to European migrants, who represented only 5 per cent of the population.[1]

In other words, it was a landgrab. Colonialism may have proceeded by different methods at different times and different places in history, but in the end it always boiled down to the same thing: a seizing of land (and the riches and labour that went with it) perpetrated by force or deception.

Two things made colonialism distinctive from other asset seizures in history. First, this landgrab was global, reaching truly planetary proportions. From 1800 to 1875, about 83,000 square miles from all over the world were acquired each year by European colonisers. From 1875 to 1914, that figure jumped to 240,000 square miles per year. By the end of that period, Britain had 55 colonies, France 29, Germany 10, Portugal and the Netherlands 8 each, Italy 4, and Belgium 1.[2] Colonialism is a story not only about the Ndebele in Zimbabwe, but also about the Bororo in Brazil and the countless other peoples who witnessed the simultaneous arrival of the telegraph, the rifle and the cross – or whatever specific combination of colonial technologies, weaponry and beliefs they were colonised with. For none of them did these things bring peace and progress, only dispossession and injustice.

The second point is that colonial stories have long lives. We are not just talking about an isolated war here or the introduction of a technology there. Colonialism is a process that took centuries to unfold, and its repercussions continue to be felt. To put it differently: the historic landgrab may be over (obviously, southern Africa is no longer a colony of the British), but the impacts of the land-grab continue to reverberate. Compare present-day England and Zimbabwe, and you soon realise that, overall, the benefits have continued to accrue to the coloniser nation in the form of accumulated wealth, while the burdens have continued to accrue to the colonised in the form of poverty, violence and lack of opportunity. We are increasingly sensing an urgent need to reinterpret our past and present in the light of that colonial landgrab.

But something else today is amiss that goes beyond this necessary reckoning with the past. Colonialism lives on in another way, through a new kind of landgrab. It is still new, but we can already sense how it could reshape our present and our future just as significantly as the old one.

This latest seizure entails not the grabbing of land, but the grabbing of data – data that is potentially as valuable as land, because it provides access to a priceless resource: the intimacy of our daily lives, as a new source of value. Is the exploitation of human life an entirely new phenomenon? Of course not. But this new resource grab should concern us because it exhibits some very colonial characteristics. It is global: nowhere is human life safe from this form of exploitation. It is very large-scale: the worldwide users of Facebook, YouTube, WhatsApp and Instagram each exceed the individual populations of China and India, the world's largest countries, with the Chinese platforms WeChat and TikTok not coming far behind.[3] It is creating unprecedented wealth based on extraction: Big Tech companies are among the wealthiest in the world (for instance, with its stock market value of US $2.9 trillion,[4] Apple is bigger than the *entire* stock market of any country in the world except the US and Japan).[5] It is shaping the very structure of the world's communications, with experts worried that the world's two largest data powers, the US and China, are increasingly associated with exclusive networks of undersea communications cables.[6] And most importantly, it continues the legacy of dispossession and injustice started by colonialism.

This book is the story of this data grab, and why it represents a reshaping of the world's resources that is worthy of comparison to historical colonialism's landgrab. It is the story, in other words, of a *data* colonialism that superimposes a data grab over the historical landgrab.[7] We already know how the colonial story develops. To get a preview of the kinds of long-term impacts data colonialism will likely have, we don't need to engage in hypotheticals. We need only to look at the historical record. Our present and not just our past is irredeemably colonial, and the new data colonialism is a core part of that.

The Four X's of Colonialism

Today, you are the King of England. But just as easily you could have ruled over colonial Spain, France or the Netherlands. Regardless of your choice, the task ahead of you is essentially the same: there are territories to be settled, resources to be traded, cities to be built, and native populations to be pacified. A fair amount of ambition and greed seem to be a requirement for the job.

Explore, expand, exploit and exterminate – the tools of your trade. With a few clicks, you apply these strategies in succession as you establish your empire. Then you apply them again. And again. And if your empire should fall, throttled by the competition or vanquished in a war, that's not a problem. You can simply start anew, because this is only a videogame: *Sid Meier's Colonization*, a turn-based strategy game released in 1994 (re-issued in 2008).

Explore. Expand. Exploit. Exterminate. This is the time-tested 'Four-X' formula for playing strategy video games. But it is also a fair summary of the formula applied by European colonisers to create vast fortunes for themselves, vast misery for everyone else, and in the process reshape completely the organisation of the world's resources.

Colonialism was a complicated project that required complicated enterprises. We've mentioned the British South Africa Company already, and there was of course the East India Company as well. The Spanish had the Casa de la Contratación de las Indias, while the Portuguese founded the Companhia do Commércio da Índia. The Dutch had their own Verenigde Oostindische Compagnie, which during the seventeenth and eighteenth centuries employed over a million Europeans to work in Asia, exporting 2.5 million tons of goods, and which was legally sanctioned to declare war, engage in piracy, establish colonies and coin money.[8] All of these companies had close links with their respective nation's rulers, and complex bureaucracies emerged around them.

In their operations, they neatly followed the Four-X model. They explored the world by launching missions to 'discover' new places they could control through military and technological means; they expanded their dominions by establishing colonies where native labour and resources could be appropriated by force; they exploited those colonies by setting up a global system of trade where those resources could be converted into wealth, always to the advantage of the coloniser; and they exterminated any opposition by the colonised, in the process eliminating their ways of being in the world. From 1492 to about the middle of the twentieth century, that's the story of colonialism in a nutshell. By applying the Four-X model, European colonisers managed to control over 84 per cent of the globe, even though Europe represents only 8 per cent of the planet's landmass.[9]

Let's see how this matches the actions of Big Tech corporations.

Today, Big Tech's efforts to explore and expand don't involve continental land, but the virtual territories of our datafied lives: our shopping habits, for sure, but also our interactions with family, friends, lovers and co-workers, the space of our homes, the space of our towns, our hobbies and entertainment, our workouts, our political discussions, our health records, our commutes, our studies, and on and on. There is hardly a territory or activity that is beyond this kind of colonisation, and there is hardly a corner of the world that remains untouched by its technologies and platforms.

But, as with historical colonialism, that territorial capture is just the start. Once colonies were established, a system was put in place for the continuous extraction of resources from these territories, and for the transformation of these resources into riches. Big Tech has achieved a similar feat of exploitation by setting up business models that convert 'our' data – that is, data resulting from tracking our lives and those of others – into wealth and power for them (but not for us). At the micro level, this means that our data is used to target us individually through advertising or profiling. At the macro level, this means that our data is aggregated and used to make decisions or predictions impacting large groups of people, such as the training of an algorithm to discriminate based on race, gender, economic status or medical condition. This is possible thanks to a rearrangement of many aspects of our daily life in such a way that ensures we are continuously generating data.

Which brings us to the fourth 'X', where the picture is more complex. In history, colonial extermination took many forms. Principally, there were deaths caused by war, mass suicide, disease, starvation and other forms of violence: 175 million indigenous people in the Americas at the hands of the Spanish, Portuguese, British and US; 100 million in India at the hands of the British; 36 million Africans who perished in transit during the transatlantic slave trade (in addition to those who perished as slaves once they arrived); one million in Algeria killed by the French; hundreds of thousands in Indonesia killed by the Dutch; and millions more who cannot be easily counted.[10]

But brutal physical violence was not the only option. Early on, colonisers realised they needed to be able to deploy other forms of extermination that eliminated not just individual lives, but also the economic and social alternatives to colonialism (which in itself entailed the extermination of life, but at a slower rate). One strategy was the imposition of agricultural monocultures that were highly profitable for the coloniser but destroyed the ability of the colonised to feed themselves. Think of the Dutch investment in coffee production in the East Indies, which went from a harvest of one hundred pounds (45 kilograms) in 1711 to twelve million pounds (5.4 million kilograms) in 1723.[11] Or think of the colonial sugar trade, which created great poverty and misery in the Caribbean while contributing a significant 5 per cent to the British gross domestic product at its peak during the eighteenth century (without slavery, sugar would just have been too expensive for most British people to consume).[12]

Another strategy of (economic) extermination was the throttling of business opportunities through the flooding of markets with cheap goods that eliminated homegrown industries. An example of this is the British cotton trade, which inundated global markets with cheap machine-made textiles that destroyed the lifestyles and livelihoods of domestic cultivators, spinners and weavers in colonies such as India; not to mention the devastating human cost paid by plantation slaves in America.[13] Throughout the colonial world, instructions like the following (sent from London to the governor of Quebec in 1763) were issued with the goal of retarding local industry: 'it is Our Express Will and Pleasure, that you do not, upon any Pretence whatever ... give your Assent to any Law or Laws for setting up any Manufactures ... which are hurtful and prejudicial to this Kingdom.'[14]

The monopolistic and anti-competitive practices of Big Tech are also having disruptive effects. The scale on which they operate cannot be ignored: if as late as 1945 one in three people on the planet was living under colonial rule, today, around one in three people on the planet has a Facebook account, and almost everyone uses search engines of some sort. The contexts and impacts are obviously different, but this resemblance in scale means that companies like Meta – which now owns Facebook, Instagram and WhatsApp – or OpenAI have a lot of power over the lives of a lot of people. Meta's power, many have argued, has contributed to the spread of misinformation and hate amidst genocidal violence (like in Myanmar), health crises (anti-vaccine disinformation) and political interference (the Cambridge Analytica scandal). Meanwhile, Sam Altman, the CEO of OpenAI, believes that the opportunity to solve humanity's problems with Artificial Intelligence is so appealing that it is worth the risk of destroying the world as we know it.[15] In other words, if AI ends up massively disrupting social values and institutions, as many experts claim could happen, Altman thinks it will be worth paying this price because of the problems AI will solve in the process. But others are not so sure this is a good bargain, which is why they are asking questions about where Big Tech's new power to determine what is relevant, normal, acceptable or true is heading.

Forms of economic and cultural extermination will, necessarily, take time to unfold, but they are the potential consequences of a change we can already see: a major shift in power relations that flows from the capture of virtual territories. Meanwhile, a very different story is being told about data, told with a much more positive twist. And here too there is a historical parallel. Colonialism has always required a strong civilising mission, an imposed worldview that dismissed all alternatives and rendered invisible all contributions emanating from the colonised. This worldview allowed the colonisers to control not just bodies, but hearts and minds as well. In the past, Christianity and Western science were the cornerstones of this civilising mission. They delineated the path towards the salvation of colonised souls and promised them a share in humanity's scientific progress, provided they remained within their assigned roles.

As an example of a historic corporation engaged in a civilising mission, consider the New England Company, also known as the Society for the Propagation of the Gospel in New England. Founded in 1649, the company engaged in the 'education' of the 'infidel' populations in British colonies, carried out through Anglican missionary work, the establishment of schools and the translation of the Christian bible into native languages. Its first president was the scientist Robert Boyle, best known for the law carrying his name that describes the relationship between the pressure and volume of a gas, but less well known for his theories on the nature of skin colour and the origins of blackness. While Boyle advocated for better treatment of slaves who converted to Christianity, he and his contemporaries never rejected slavery wholesale, and as members of the Royal Society they benefitted directly from the slave trade through that institution's investment in Royal African Company stock.[16]

Big Tech too has a civilising mission that is mixed up with its technologies and business goals. Part of this civilising mission continues to revolve around Western science: network science, data science, computer science, and so on. The other part no longer revolves around Christianity, but around parallel sublime notions like the convenience that will supposedly make all our lives easier, the connectivity that apparently will bring new forms of community, and the new forms of science and Artificial Intelligence associated with machines that purportedly can solve problems better than humans. It's not as if some of these dreams are not becoming real for a select few; it's just that they risk becoming nightmares for everyone else in the form of lost livelihoods, new forms of exploited labour and the loss of control over vital personal data.

Civilising missions, economic motives, the exercise of power and the introduction of specific technologies have been deeply intermingled throughout the history of colonialism, but always with an uneven impact that favours some but not others. We saw this already in our example of the telegraph in southern Africa.

Another example was the introduction of the electrical grid to India throughout the Madras Presidency in the early twentieth century. Electricity was considered a triumph of Western science over the 'devil of darkness', and while it was initially used exclusively to improve the lives of white people as a display of cultural superiority, its application was eventually extended to the rest of the population as a kind of advertisement for the supposed benefits of colonialism. It powered cinemas, illuminated public spaces, propelled tramcars and provided energy to places like hospitals – all while generating income for British companies. But beyond these comforts, amusements and public services, electricity also served to run the lighthouses that guided ships carrying colonial goods, powered weapon factories and electrified prison barbed fences that kept the population in check, extended the operating hours of offices and printing presses carrying out the coloniser's administrative work, increased revenue by accelerating industrial and agricultural production, and provided the backbone for communication and transportation networks that guaranteed the smooth functioning of the empire. In other words, behind the civilising mission of this 'gift' from the coloniser, electricity was instrumental in sustaining the core business of colonialism,[17] which was anything but peaceful (100 million Indians were exterminated during British rule, as we mentioned earlier).

Replace 'electricity' with 'data' and, while the specifics are different, some elements of the story remain eerily similar. Ways of processing data are also heralded as scientific achievements, a gift that promises convenience, connectivity and new forms of intelligence. But look under the surface of this civilising gift, and you will find that it also brings new forms of surveillance (through facial recognition or workplace monitoring), discrimination (when algorithms deny or control access to services based on people's profiles) and exploitation (when gig workers' wages are continuously adjusted downwards, for instance).

A discussion of the colonial legacy of Western science will be a recurring theme throughout the book, and this is a touchy subject. To point out the ways in which Western science has been used to justify social and environmental harms might come across as a wholesale dismissal of the benefits and contributions of science, which are many (not least to monitor and model the harms that humanity is currently doing to our environment and, if we can find them, monitor potential solutions). In no way is our argument anti-science, nor do we wish to fan the flames of science denialism. But that doesn't exempt us from facing head-on the important critiques that colonised peoples have made of the ways in which Western science was used during and after colonialism to control and exploit the natural and social realms. In fact, it is only by looking at contemporary science through this colonial lens that we see these continuities, which go back to the origins of modernity generally and of modern science. That is all the more vital when this problematic legacy continues to shape developments like data science and AI, which have huge impacts on our present and our future. It is exactly as the Cameroonian philosopher Achille Mbembe has said: 'Our era is attempting to bring back into fashion the old myth that the West alone has a monopoly on the future.'[18]

Terms and Conditions

Our new colonial reality entails new processes of data extraction that are changing education, healthcare, agriculture, policing, the way we talk to each other and the way we judge each other, and many other areas besides. Which is why this latest stage in colonialism's development requires a new term: *data colonialism*, a social order in which the continuous extraction of data from our lives generates massive wealth and inequality on a global scale.

This new social order comes with a new social contract, based on the premise that the data 'exhaust' we generate through our online interactions should be given to corporations for free. Why? Because, we are told, the data we generate can only be processed or refined using vast computing and storage capacities that we do not have at our disposal. Humanity's progress, the story goes, depends on that surrender of data. And we are told that this surrender is happening with our full consent, because after all, didn't we click the 'I accept' button when installing those apps on our phones or using those platforms?

Let's pause for a moment and consider what happens when we click 'I accept'. Or rather, what happened if you clicked 'accept' in 2007 under Google Chrome's Terms of Service (TOS) agreement. Most people don't read the Terms of Service in detail. So it's worth setting out some of what you would have agreed to back in 2007 when installing the web browser:

> You give Google a perpetual, irrevocable, worldwide, royalty-free, and non-exclusive license to reproduce, adapt, modify, translate, publish, publicly perform, publicly display and distribute any Content which you submit, post or display on or through, the Services.[19]

More recent versions are not worded as rapaciously. They don't need to be! Because, without most of us ever having read those original terms and conditions, we have reorganised our lives around whatever rules Google imposes on us. Google's new rules can afford to be worded more softly because the extraction of data is already well under way – blunt language is no longer necessary. In any event, on July 2023 Google updated its privacy policy to indicate that it reserves the right to 'use publicly available information to help train Google's AI models and build products and features like

Google Translate, Bard, and Cloud AI capabilities.'[20] This means that if any of us post anything online publicly, Google has unilaterally decided it can use it for training its AI models, without asking anyone for permission.

Now, compare Chrome's original TOS to a much older document, the *Requerimiento* (which in Spanish means the 'Requirement' or 'Demand') written in 1513, also quite early in the first colonial landgrab. This was a document read by the Spanish conquistadors as they entered a village or city in the Americas in the middle of the night to wreak havoc and get their pillaging under way. They read this document in Spanish . . . to indigenous or First Nation[21] peoples who did not speak Spanish! No amount of translation would probably have been enough to explain the differences in worldviews to indigenous peoples as they heard these words being uttered:

> But, if you do not [submit to Spanish rule], I certify to you that, with the help of God, we shall powerfully enter into your country, and shall make war against you in all ways and manners that we can, and shall subject you to the yoke and obedience of the Church and of their Highnesses; we shall take you and your wives and your children, and shall make slaves of them, and as such shall sell and dispose of them as their Highnesses may command; and we shall take away your goods, and shall do you all the mischief and damage that we can . . .[22]

We are not suggesting that what followed the reading of the *Requerimiento* is the same as what follows our installation of Chrome. Google won't enslave us or our children, and certainly the cost of rejecting Chrome's terms is not extermination. But the point here is to highlight the tricks and threats necessary to perform such initial acts of dispossession. Incomprehensible language must be used, because extraction requires that we don't fully understand the terms under which it is being carried out. By the time we do, it is too late.

The Spaniards, and all subsequent colonisers, counted on this initial failure to understand what was behind the act of dispossession. They knew the *Requerimiento* was a pretence, a feel-good excuse to legitimise in legal, political and religious terms the atrocities they were about to commit. Indigenous people quickly sensed it, too. They would flee instead of waiting around to hear the proclamation; they had to be captured and tied to trees to make them sit through its reading, or they had to be enticed with cheap gifts. In one recorded instance, a chief of the Cenú people insisted that the whole thing be explained to him, and when he finally managed to grasp the basic meaning, he said that a Pope who was giving away land that was not his must be a drunk, and a king who accepted this gift must be a lunatic. Bartolomé de Las Casas, a priest sympathetic to the indigenous people's cause, said that he didn't know whether to laugh or to cry upon hearing the *Requerimiento*.[23]

Raw Materials

Perhaps one day we too will wonder whether to laugh or cry when reading the TOS of our present-day apps which carry much of Google Chrome's original intent. In the meantime, the dispossession of our data proceeds largely unabated. While it is the Big Tech companies that get the critical headlines, the data grab is not just down to a few rogue companies: it is happening at every scale, sometimes in dark corners and sometimes in plain sight.

Take for example Lasso, a leading marketer in the US health sector who you almost certainly have never heard of. Personal health data is widely assumed to be legally protected, but Lasso has found a way to offer a number of products for marketers who want to reach customers interested in healthcare, including Blueprint™, Connect™, and Triggers™. While Lasso, as one would expect, says it is compliant with US health data regulations, its ambitions for Blueprint™ are striking. To quote from its webpages:

> Lasso Blueprint enables marketers to create high-value audiences composed of health providers and consumers based on *diagnoses, medications, procedures, insurance data, demographic information, and much more.* The product provides audience counts in real-time ... audiences can also be dynamically refreshed on a weekly basis with the latest real-world data to ensure you [the marketer] never miss an opportunity to engage with your targets.[24]

So, even if your health data remains strictly anonymised, it is collected in bundles which are then used to uniquely target you as a consumer with a particular medical history. It hardly matters whether your name is on the data package. All this relies on sophisticated data processing by Xandr, a company that was acquired for US $1 billion by Microsoft in 2022, and operates far from news headlines.

Data capture through surveillance for the purpose of marketing or algorithmic profiling was the biggest problem we had to worry about until recently. But the vertiginous evolution of AI in the last few years has proven that data colonialism is unleashing effects that may transform how we think or create – or, more specifically, how we allow machines to do thinking or creating for us. Consider what is happening in the art, culture and media sectors with the arrival of natural language generation and generative AI tools. These programs are mimicking our creative endeavours using Artificial Intelligence algorithms, with increasing degrees of authenticity. Given a few text instructions, AI tools like ChatGPT, DALL-E, DeepDream or DeepMind (the first two owned by OpenAI, the other two by Google) can generate text, images, sound or human speech that not only seem like they were generated by a human, but can imitate a specific author like Jane Austen, a painter like Salvador Dalí, or a musician like Fela Kuti.

For all this to be possible – for the AI to copy what Austen, Dalí or Kuti read, paint or sound like – it needs to learn what other artists, or indeed all of us, read, paint or sound like. In other words, it needs to analyse not only Austen's novels, but other novelists' works too, as well as what you have written; it needs to analyse not only Kuti's voice, but your voice too. It can fetch them from repositories that may contain the videos we upload to social media platforms, or the voice messages we leave on our friends' phones.

Some people may not think the AI generation of derivatives that mimic someone famous or non-famous are that big a deal. They can be seen as an entertaining gimmick, or perhaps a useful work tool (imagine being able to edit a voice or video message without having to re-record it). But questions about authenticity, about the value of original work, about our ability to recognise altered records, and about who controls the power to perform these feats are worthy of our consideration in these early days of Generative AI.

And there's a bigger point too: that our collective cultural and social products now serve as the extracted raw material on which AI relies. For example, Google's MusicLM software, which can generate melodies based on instructions like 'Meditative song, calming and soothing, with flutes and guitars,' was trained on 280,000 hours of music. Did Google pay to license all of the music it used for this purpose? Not likely. This is probably the reason why it decided to delay the release of this tool for a while. If the generated music were to sound too much like the source material from which it derived, this would open the door to potential lawsuits.[25] But eventually Google released the program free of charge, like most of its products. In its final form, the AI will not comply with requests to copy specific artists or vocals, allowing Google to avoid potential charges of copyright infringement. But there is still a corporation expropriating our cultural production as source material to train a machine to do the work of humans, because machines will be able to do the work more quickly and cheaply.[26]

From internet searches to cloud services to generative AI, Google doesn't need to charge us to use its products, because *we are the source material for its products*. Allowing the public to use its tools for free (as a way to 'empower the creative process', in the case of MusicLM)[27] cannot conceal the nature and scale of this data 'heist', as Naomi Klein has called it.[28]

Whether hidden in dark corners or not, these are extensive acts of appropriation. We are talking about the capturing and monetisation, through data, of our collective activities, our interactions with each other across time and space, our shared resources. The 'cool' factor of 'generative AI' is basically a device to distract from this.

Welcome to data colonialism. It is happening everywhere. It is an appropriation of resources on a truly colonial scale. A data grab that will change the course of history, just as the original colonial landgrab did five centuries ago.

Reading the Present through a Colonial Lens

Comparisons between the old colonialism and the new colonialism are worth exploring, but to compare is not necessarily to equate. History is too complex and interwoven for that. There is therefore no almanac we can offer you that would dispense simple one-to-one comparisons between, say, the Dutch Verenigde Oostindische Compagnie and Alphabet-Google. How could history repeat itself in such a simple way? Instead, this book will offer you a *lens* through which to see that today's data grab really is a continuation of the colonial landgrab that reorganised the control of the world's resources starting five centuries ago, putting the opportunities for exploiting them in the hands of the few. This lens enables us to identify six important similarities between historical colonialism and data colonialism.

First, and most fundamentally, both systems are founded on an appropriation of the world's resources, treating them as 'just there' for the taking. Historically it was the territories, natural resources and labour of the Americas, Africa and much of Asia and the Pacific that were plundered. In the contemporary era, it is human life, extracted in the form of data, that is being appropriated.

Second, that appropriation always serves a wider purpose: to build a new social and economic order that locks in that landgrab of resources. Whether this took centuries to develop (like in historical colonialism) or mere decades (like in data colonialism), the goal has always been to construct a new way of ordering all aspects of life to benefit the coloniser by locking in a form of extraction.

Third, this colonial order and global landgrab of resources is the joint work of states and corporations. In the past, colonial states were monarchies, and the companies licensed by them (like the British South Africa Company) were special-purpose vehicles created for colonial extraction. Today, corporations have played a much more important role in building complex alliances with nation-states and their governments, as we can see in the example of China and India, but also in different ways in the US and Europe. But the fact remains that both historical and data colonialism represent state-corporate partnerships.

Fourth, colonialism has always had a disastrous effect on the physical environment, and continues to do so. While global warming began its ascent during industrial capitalism, it was colonialism that set the stage for this by framing the natural world as something cheap and disposable whose only value was the profit that could be extracted from it. Some forms of environmental devastation are as old as colonialism, like the large-scale mining of precious materials which the electronics sector still relies on. Others, while more recent, have strong links to earlier periods of colonialism: think of the devastation enacted today by industry sectors like Big Oil and Big Agriculture, which rely on the control of original colonial territories. And others are entirely new effects of data colonialism; for example, energy and water consumption by data centres, carbon emissions associated with online shopping deliveries, or the increased number of satellites orbiting the planet.

Fifth, and very importantly, colonialism has always generated deep inequalities that divide extractive elites from large populations of exploited labour. In fact, the very idea just mentioned that colonial land was 'just there' for the taking was based, from the very start, on assuming that the people who were already living on colonised land were worth nothing. Data colonialism introduces new forms of inequality connected to the exploitation of data, which often replay and deepen those older inequalities, including racial and gender discrimination. And all of the developments with data extraction that we'll describe in later chapters are played out in a highly unequal global economy and geopolitics that have been profoundly shaped by colonialism.

Finally, the predatory acts of colonialism are so manifestly unjust that they have always had to be disguised by more positive 'civilising' narratives, or 'alibis'. In past centuries, those narratives were tied to Eurocentric notions of progress or Christian ideals of salvation. In the Big Tech era those narratives connect more closely to the values of a capitalist economy, while retaining echoes of earlier civilising missions. Can we imagine Facebook taking on the role of providing internet infrastructure to over thirty African countries without these sorts of narratives?[29]

Despite these six similarities, there is no denying that there are significant differences between historical colonialism and data colonialism. We are, after all, talking about data, not land, which is a different type of asset. While there are limits to the resources that can be extracted from a single piece of land, the same data set can be exploited ad infinitum by an unlimited number of parties, which is what economists mean by saying that data, unlike land, is a non-rival good.

And yet, while both historical colonialism and data colonialism involve acts of extraction on a vast scale, we must remember that extraction only marks the first move, not the last one, in the building of a social and economic order, and that this order is shaped inevitably by the levels of organisational complexity typical of its time (without doubt, our economy and its organisational structures are massively more developed today than in the sixteenth or eighteenth centuries).

Rather than talking about data colonialism in terms of similar or different effects, it is therefore more accurate to talk about *compounded* ones. The profound injustices of historical colonialism (theft, slavery, discrimination), far from being superseded by the new injustices of data colonialism, are made worse by them. This is an important point, because our use of the terms 'historical colonialism' and 'data colonialism' might give the incorrect impression that, as authors, we believe one kind of colonialism is over and a new one beginning. However, when we say that data colonialism is a new form of colonialism, we wish to convey a sense of evolution, not of spontaneous generation. Thus, data colonialism should be considered on a par with other terms like plantation colonialism or settler colonialism. All of them describe different yet overlapping aspects and impacts of a phenomenon with a long and ongoing history: colonialism.

Our emphasis on the dangers of data extractivism might also give the impression that we are against data altogether, and are denying there are ever good reasons for collecting and processing it. That is not our position. There are of course many valid reasons to collect and use data, especially when the purpose is to understand or even change the world for the better. What we are against is the use of data for the exclusive purpose of generating profit without social validation and public control, especially when this is presented to us by corporations as a narrative of salvation and progress. But why think that all good uses of data must come attached to the extractive terms and conditions that we'll discuss throughout this book? It is not data, but those terms and conditions, which currently define and shape the world of Big Tech and AI. We must contest and rethink those terms and conditions, and the lens of colonial history can help us do that.

Your Guide to the Book

We are aware that colonialism is a difficult subject. People come to this subject from different histories and standpoints, including different relations to the history and reality of colonialism. Because of this, we acknowledge that people will inevitably be more or less ready to hear our argument. We respect that, because as authors we too are differently positioned in relation to colonialism, one of us having grown up in London and the other in Mexico.

In Chapter One ('A New Colonialism') we'll confront the doubts many people might have about using the lens of colonialism to explain what is happening with data today. In Chapter Two ('Data Territories') we'll lay out in more detail what sorts of data are being taken, how it is processed, and what sectors of life are being affected. We'll find that it's pretty much every sector of the economy, society and government, which raises some important questions about how power relations in contemporary societies are being transformed. In Chapter Three ('Data's New Civilising Mission') we look in more detail at the stories we are told (about data extraction and how this benefits us) that effectively disguise the realities of data colonialism. In Chapter Four ('The New Colonial Class') we'll ask who is seizing this data, surveying the sorts of corporations involved, the motivations of those who work for them, and the individual entrepreneurs who are working at the data frontier.

In the book's last two chapters we think about resistance. If the cruel and divisive legacy of historical colonialism still lives on, it is hard to imagine what it would take to resist the new stage of colonialism that is unfolding. But imagination and resistance are connected. We can think more sharply about what resistance to data colonialism might look like by drawing, as we do in Chapter 5 ('Voices of Defiance'), on inspirational figures who opposed historical colonialism in the past, as well as figures in the early history of computing who predicted the dangerous tendencies of today's digital world. Then, in Chapter 6 ('A Playbook for Resistance') we turn to the practicalities of actual resistance, drawing on examples that have been emerging around the world.

The world's information sources are today being carved up between corporations, often with national governments in tow, just as were the lands of historical colonies. Although the outcomes will, as always in colonialism, be unequally distributed, this new data grab will affect the living conditions of everyone on the planet in one way or another. This data grab is happening right now. It is neither an accident, nor the work of a few rogue capitalists. And it's a global process, even if the techniques and civilising missions vary in detail around the world.

Yes, data extraction often seems highly technical and remote from our daily lives. But in its effects, it's anything but remote, which is why we'll start each chapter with an individual story. While the people in these narratives are fictional (the impacts of data extraction are everywhere, but the individual stories tend not be recorded), they are composites that are always grounded in real facts.

Taken together, the book's chapters provide a map to make sense of this unfamiliar and disturbing new landscape of data extraction. At the end of our journey across this landscape, we'll look back and ask: can we *really* afford to do nothing by way of resistance?

CHAPTER 1

A NEW COLONIALISM

AS A SINGLE mother in the US, Tracy cannot imagine what she would do without technology, and considers herself lucky to be living in our Digital Age. Her phone and laptop are with her day and night, and she uses free apps to schedule doctor appointments, order food and supplies, pay bills, receive money from her parents, coordinate childcare, communicate with her daughter's teachers, and so on. In the occasional moment of peace and quiet, she uses these apps to entertain herself and socialise with friends and family.

Tracy took a course in college where the professor mentioned some of the privacy threats that come with using these services, but she remembers thinking that she didn't have a problem with corporations collecting her data, because she was not doing anything wrong or worth hiding. Tracy is not a terrorist, so why would she care if a company records information about her in exchange for the opportunity to use their services for free? Sure, the ad-targeting could feel a bit like stalking (sometimes it seemed like her phone was listening to her and then showing her ads for things she might have mentioned in conversations). But this was all just part of our capitalist system which, while not perfect, was still the best option available.

Tracy's views began to change when the Supreme Court of the United States overturned *Roe v. Wade*, opening the door for

US states to criminalise abortion. As someone who had had an abortion, Tracy thought this was an important issue, and she became concerned when she started reading about how some of the data collected by the apps and services she used could potentially be turned against her. Articles were telling her to delete her period-tracking app and stop using her fitness monitor. A friend mentioned that her smart water meter, installed in her apartment two years ago, could determine if she was pregnant – or stopped being pregnant – just by analysing the changes in her household's water consumption patterns. 'Think about it,' her friend told her: smart water meters can detect when and how often the toilet is flushed. Since pregnant women tend to use the toilet more frequently, a sudden increase in toilet use might indicate a pregnancy, while a sudden decrease might indicate an abortion has taken place. Tracy also knew her shopping apps could make a pretty accurate guess about what was happening in her uterus based on what she purchased, and she'd also read stories about women who claimed Facebook knew they were expecting before they did.

Suddenly, Tracy went from thinking of herself as someone who was not doing anything wrong to feeling like someone whose data could be used to prove she *had* done something 'wrong', that is, ending a pregnancy. True, none of that had happened in the state where she lived yet, and hopefully it wouldn't come to pass. But she read on the news that Facebook gave police in the state of Nebraska access to a teen's direct messages to prosecute her for an abortion, and that lawmakers in Texas were working on a bill that would make it illegal to provide information on how to access abortion services. If the bill passed, some people feared social media platforms would start to censor posts about abortion.

Tracy found herself wondering what exactly she had consented to every time she clicked the 'I agree' button on the dozens of apps she had installed. What data was being collected about her, and how could it be used? Why was it being collected, and who in the end would benefit? She couldn't find the answers to those questions easily, not even by googling. Sure, Google had just announced that they wouldn't record the location data of women visiting abortion clinics, but this implied that they were collecting data about all the other places women were visiting (she later read that, despites its promises, Google was still tracking visits to abortion clinics[1]).

The fact that these services were free no longer seemed like such a good deal to Tracy, after all. With only a tiny turn of the political dial, something previously considered normal could become illegal, and Tracy's data could suddenly implicate her.

Many of us know someone – or are someone – in a similar position to this person we are calling Tracy. But could we say that Tracy is a colonial subject? Slaves were colonial subjects. Oppressed and disenfranchised people serving vast empires hundreds of years ago were colonial subjects. Tracy doesn't seem to be any of those things. In fact, apart from some data privacy concerns, she seems to be living a relatively comfortable life and enjoying the convenience afforded by digital devices. Yes, it's scary that her personal data could be used to target her in some way. But isn't it bad governments, not good corporations, who are to blame for that kind of persecution? Isn't it therefore irresponsible, even insulting, to suggest that Tracy is the victim of a new colonialism?

We're not claiming that digital platforms take us back to the brutal colonial life of past centuries, with its continental-scale forced labour and inhuman living conditions.

Our argument is a different one: that our digital lives today, and the kind of relationships we have with the corporations which make those lives possible, involve fundamental changes in power relations that can only be fully understood within the framework of colonialism, and the profound change in the world's resource distribution that colonialism engendered.

No Capitalism without Colonialism

In order to see this unfolding reality we need to look at the colonial past and the capitalist present in a different way. No one doubts that we live today under an arrangement for managing the world's resources called capitalism, an arrangement that organises economies through markets. By suggesting that a new colonialism is emerging in the midst of this order, we are establishing a new kind of relationship between colonialism and capitalism.

Before going further, it might be helpful to start with some basic definitions of what colonialism and capitalism are. According to Wikipedia:

> Colonialism is a practice or policy of control by one people or power over other people or areas, often by establishing colonies and generally with the aim of economic dominance. In the process of colonisation, colonisers may impose their religion, language, economics, and other cultural practices. The foreign administrators rule the territory in pursuit of their interests, seeking to benefit from the colonised region's people and resources.[2]

While Wikipedia's page for capitalism reads:

> Capitalism is an economic system based on the private ownership of the means of production and their operation for profit . . . In a capitalist market economy, decision-making and investments are determined by owners of wealth, property, or ability to maneuver capital or production ability in capital and financial markets – whereas prices and the distribution of goods and services are mainly determined by competition in goods and services markets.[3]

Judging from these definitions, it would seem that colonialism and capitalism are two completely different things.

If we follow Wikipedia, colonialism is a practice (or policy), whereas capitalism is a system. More to the point, colonialism seems exploitative, taking things away from people through force, exerting power from a distance to make sure, among other things, that the colonised accept and adopt the ways of thinking and acting of the coloniser. In contrast, capitalism doesn't seem as brutal. It's almost civilised, a matter of arranging resources and making better economic decisions. Yes, there still seems to be one group of people in control (the owners of the means of production), but there is no mention of violence or force. Instead, we have markets where capitalists compete, which presumably prevents abuse and creates benefits for the entire economy.

We know these definitions are incomplete and, depending on where you stand, controversial. But whether you accept their detail or not, they capture the common-sense view that colonialism was about the exploitative *extraction* of things, whereas capitalism is about the collaborative *production* of things. Those tenses ('is', 'was') are not accidental, because in the standard view of history, colonialism came *before* capitalism. In fact, the standard view of history would tell us that the brutality of colonialism needed to end in order to make room for the more civilised mode of production that we all know as capitalism. We must question that standard view of history.

Capitalism Cannot Be Understood without Colonialism

There is no way to understand the capitalism that Tracy recognises as her daily reality without also understanding the role colonialism played and still plays in making it possible. We need to look beyond our immediate historical context and recall that the wealth generated in the mines and plantations of colonialism financed the factories of early industrial capitalism.

Sometimes we can track the transformation of colonial wealth into capitalist resources quite directly and within one generation. Take the case of pro-slavery parliamentarian Richard Pennant, 1st Baron Penrhyn, who owned six sugar plantations and hundreds of slaves in Jamaica while being actively involved in the development of the Welsh slate industry. Or the case of Scottish entrepreneur John Pender, who made his fortune manufacturing goods from cotton produced in slave plantations in the US, and then invested much of his wealth laying undersea telegraph cables connecting Britain to America and India. Of course, the actual transfer of wealth was often more complicated than that, involving multiple parties, institutions and nations. But the fact remains that colonialism greatly facilitated Europe's transition from feudalism to capitalism.

How exactly? According to one estimate, in the last 300 years colonialism contributed to an increase of 14 to 78 per cent of the total average income for the coloniser nations. Furthermore, it was colonialism that allowed European protocapitalists (merchants, artisans, landlords, etc.) to clear peasants off the land, accumulate capital, increase production and hire wage workers quickly, all with the support of their respective states. This accumulation of wealth and power allowed Europeans to triumph over competing proto-capitalists in Africa, Asia and China, cementing Europe's global dominance and allowing it to re-invest its wealth into additional colonial enterprises. Without the accumulated wealth of extraction, business entrepreneurs wouldn't have had the stable resources to own the means of production and so become capitalists.[4]

In any case, it makes no sense to think of colonialism as something primitive that had to be got out of the way before capitalism could develop. Karl Marx wrote that 'the overseer's book of penalties replaces the slave-driver' under capitalism, but he was wrong.[5] Slavery, a major way of organising labour on a global scale under colonialism, thrived during the period of industrial capitalism's growth (the nineteenth century). This was, for example, the time of the massive expansion of slave plantations in the US South. Meanwhile, the UK's stance of outlawing slavery hardly prevented the UK economy (for example, the port cities of Bristol and Liverpool) from benefitting hugely from the economic outputs of slavery. Economists Thomas Piketty and Gabriel Zucman argue that in the nineteenth-century US, the value derived from trading and owning slaves was more than the wealth derived from factories and transport combined. Contemporary historians have increasingly come to realise that, far from being organisationally backward, colonial production methods were at the cutting edge of nineteenth-century accounting, financing and the 'rational' management of human bodies to produce economic output on the largest possible scale and at the lowest possible cost.[6]

We can take this argument even further. There's no way the huge wealth concentration in the Global North today, including the US's dominant position in the global economy, can be understood apart from the unequal historical legacy of nineteenth- and twentieth-century empires. That legacy is the source of the continuing uneven opportunities open to the world's national economies: the 'New International Division of Labour', which involves the shifting of low-price manufacturing out of much of the Global North into the Global South;[7] and the whole structure of multilateral organisations like the World Bank and International Monetary Fund, dominated by the West, that manages the so-called 'development' process.[8]

The standard view of history interprets the past five centuries of history through just one lens at a time, when in fact we always need two, superimposed lenses: colonialism and capitalism. It is misleading to see colonialism and capitalism as discrete historical periods linked only through a process of historical evolution, with colonialism (whose brutality has fortunately ended) coming first, and capitalism (under which we are happily living) coming later. Colonialism has been there all along, first laying the foundations of a future capitalism and then fuelling an insurgent industrial capitalism in the growing global economy. Capitalism has always had a colonial gene within it. And that remains true, even though, in much of the world, there are no more imperial colonies (although the peoples of Puerto Rico, Palestine and American Samoa might have a different view on that).

To be clear, we are not saying that capitalism is just colonialism in disguise (for there really is such a thing as capitalism, and it has unique characteristics in terms of its economic and social organisation). Rather, we have to treat colonialism and capitalism as parallel and linked processes. In fact, the system of profit generation that is capitalism always depended on the colonial landgrab that started two centuries before capitalism. That colonial landgrab continued through the nineteenth and twentieth centuries (think of oil and the extraction of ever more rare minerals), and it continues today.[9] Capitalism has a colonial dimension, not by accident, but by design.

If colonialism and capitalism – extractivism and production – go hand in hand, today's new methods of 'mining' data show that similar processes are continuing to unfold. This is why writers who interpret what is happening with data exclusively in terms of the dynamics of capitalism cannot quite make sense of the scale and force of the data extraction under way today. There are multiple theories that propose we are experiencing a new form of post-industrial capitalism – data capitalism, informational capitalism, platform capitalism, or, as scholar Shoshana Zuboff has made famous, surveillance capitalism.[10] There is a lot that is useful in those analyses, as they help us see the particular ways in which, within capitalism, data is changing how we produce and consume things. But one crucial thing is missing from these theories: a historically contextualised account of the sheer scale and seemingly unstoppable social and cultural force of the data grab under way.

It is the huge transfer of the world's extracted resources throughout colonialism's history that is the crucial reference point for making sense of the global data grab. The landgrab of the world's data is not a recent aberration (a side-effect of technological development that has somehow gone wrong) as prominent critics such as Zuboff seem to suggest,[11] but a continuation of what is now a very old story of extraction. To ignore this and pretend that the great accumulation of wealth today by Big Tech doesn't have an extractivist undercurrent that mirrors earlier colonial landgrabs is to engage in historical amnesia. Acknowledging, as we must, that in many parts of the world capitalism still benefits from the legacy of historic colonial dispossession should not prevent us from seeing the new colonial landgrab that is under way before our eyes.

Only thinking through colonialism and capitalism in this way enables us to see the startling possibility implied in Tracy's worries: *Yes, there is a new colony, a new zone of extraction. Data colonialism exploits it. What it exploits is our lives as human beings. Wherever we are.* It is the lives of all human beings that are potentially for the taking. And this capture is part of our devices and platforms working exactly as intended: it is not a result of rogue activity or something going wrong. Exploitation is a feature of the routine operation of our devices.

Does this mean that today's profound inequalities in who suffers most, inherited of course from historic colonialism, no longer matter? Absolutely not. Does it mean that this new colonialism threatens something fundamental to the lives of all human beings, whatever their race, class, gender or sexuality? Yes, it does. It is this strange combination – the continuation of colonialism's inherited inequalities and colonialism's acquisition of new tools that potentially affect human life in radically new ways – that we must understand. And it is exactly this combination that the Cameroonian social theorist Achille Mbembe refers to provocatively, even for many counter-intuitively, as 'the Becoming Black of the world',[12] a phrase we'll unpack in Chapter Five.

But what does this mean for our view of colonialism as an evolving historical phenomenon?

Colonialism, Like Capitalism, Keeps Mutating

If colonialism is still relevant to how we interpret contemporary capitalism's latest practices with data, what does that mean for how we understand colonialism as a long-term historical process?

Colonialism and capitalism keep changing and adapting, like most historical forces do. Social movements, scientific understandings, cultural trends, religions . . . those things don't just 'begin' and 'end'. They are not static. They keep evolving and transforming. This is true of capitalism, and we argue that it is true of colonialism too.

The most obvious mutation in colonialism is that we've gone from a landgrab to a data grab. This is a relatively recent development, but that doesn't mean it won't alter the organisation of human life for a long time, just as the impact of the original colonial landgrab still reverberates today. The data grab is already changing how education, health and our workplaces are experienced in more and more parts of the world.

Another important mutation in colonialism is the different role that violence plays in the new iteration. Historic colonialism was brutal and inhuman. But as we saw in the last chapter, it also found ways to operate without direct physical violence (mostly through cultural and economic violence). Which brings us to an important question: if physical violence is largely absent from data colonialism, can we still call it colonialism?

Here we must keep in mind that, as we argued previously, colonialism is evolving in sync with the social relations that are part of capitalism. Centuries ago, colonisers had just two means at their disposal to grab the land and the gold: violence and lies. They often used both, killing, pillaging and exchanging trinkets for gold. At that time there were no alternative ways of achieving the colonial goal of grabbing resources at the scale, speed and profit margin the colonisers wanted, for colonisers and colonised had absolutely no pre-existing social relations with each other.

Today, after five centuries of colonialism and two centuries of capitalism, that kind of violence is no longer necessary, or rather, it has evolved and become more symbolic. Your dispossession (your loss of control over the data that affects you and the impacts this has on your ability to control the terms on which you work, get loans, educate your children, and so on) may be no less absolute, but no violence is needed to persuade you to click the box that says 'I agree to the terms and conditions' before installing an app. That click alone, by virtue of the vast legal and practical infrastructure of capitalist social relations, is enough to plunge us into endless spirals of data extraction. In other words, today's forms of extraction are almost frictionless, although that doesn't mean their long-term repercussions are entirely non-violent, as indeed Tracy feared in terms of the pregnancy app's implications for her control over her own body. The key point is that, violence or no violence, a comparable colonial act of appropriation can still be performed. This takes us to one more observation about the link between colonialism and capitalism.

Colonialism's Core Mission Stays the Same

If colonialism keeps evolving, we cannot expect its present permutations to be a mirror image of the colonial practices we've seen over the past 500 years. Indeed, it's pointless as well as offensive to suggest that using Facebook makes us slaves in the same way that plantation workers were enslaved. Of course it doesn't, nor does it need to, for a colonial-scale extraction to be going on.

Lots of things have changed in five centuries, and we would expect today's data colonialism to be distinctive in its contours from its previous versions of colonial appropriation. We can best explain this by separating out the form, content and function of colonialism throughout history.

Because colonialism has developed over centuries, its form and content have shifted and adapted with the times. By *form* we mean the social and economic features of colonialism at any one moment. Think, for instance, of how planter colonialism (grabbing land to grow cash crops, while not necessarily living on that land) was very different from settler colonialism (appropriating land to live on it). By *content* we mean the specific ways that overall form manifested itself in different places and times. Think of how colonialism in Mexico by the Spaniards looked and felt very different from colonialism in India by the British. All of these historical forms of colonialism are obviously very different, in turn, from data colonialism. In no way do we want to deny those differences.

But more important is the similarity, the thing that these forms of colonialism share, which makes them all forms of historically evolving colonialism. And that is the *function* – or, if you like, the core mission – of colonialism. The core act, the basic meaning of all historical types of colonialism, was to extract, to dispossess. That is why the image of the landgrab is still relevant to understand what's going on with data today, because colonialism always and everywhere means dispossessing others of what is theirs, and taking it regardless of their rights. Whatever the nationality of the coloniser, whatever the geographical location or time period, whatever the legal structure, colonialism has always been about colonisers taking something that does not belong to them through forceful means.

We are not merely saying that what's going on with data is like historical colonialism, or resembles it at a superficial level. We are saying that, through our relations to digital platforms, a new and expanded colonial order really is being built. This new colonial order based on data extraction will dispossess almost everyone in the long run of the ability to control the judgements that are made about them; and it will do this based on the data extracted from their lives, just as Tracy fears.

All this is so far happening without much outcry. A key reason is that we are fed narratives that disguise the reality of what's happening. One such narrative repeats a colonial view of resources as *cheap*.

The Colonial Economy of 'Cheap' Resources

Businesspeople often say that the data they extract from our time online is 'just there'. It's what we generate, effortlessly, as we move around online, and unless they (the corporations) put it to good use through careful data processing, it would go to waste. Wouldn't it be too bad if all that knowledge were lost, the data that might help solve most of humanity's problems? So why not let control of this vast 'data exhaust' pass to business for free?

The echoes here of the arguments historical colonisers once used to justify their taking of somebody else's land are uncanny. They argued that the land and natural resources they took were abundant, free, and 'just there' for the taking – in other words, cheap. There was a whole 'new' world waiting to be developed by the coloniser, whose arrival was often deemed sufficient to justify a declaration that the new territories were his and his alone. An old legal doctrine was conveniently revived to make sense of this: the land taken was said to be without owner (*terra nullius*, which in Latin means No One's Land).

Inconveniently for the coloniser, some of that land had people living on it. But from their Eurocentric perspective, those people had no rights to the land, in part because they just weren't using it to its full capacity (they weren't 'civilised' enough). Because the colonisers could be relied upon to make the land more productive (according to their definition of productivity), they had an over-riding right, they believed, to just take the land and do what they wanted with it.[13] Later colonisers, particularly the English, developed the idea that colonised peoples were of a human type that was 'naturally servile', compared with the colonisers' supposed natural propensity to rule.[14]

In order to put those natural riches to good use, the colonisers realised they needed another important resource: human labour. Thus, the labour of the colonised came also to be portrayed by the coloniser as cheap and abundant – just like cheap nature. According to the coloniser's view of the world, some people were predestined to do that work, and this fate was mostly determined by their race. The labour of local or imported populations of colour in the colonies came at a low cost, at least if you excluded the costs of the violence that held it under control. Jan Pieterszoon Coen, a governor of the Dutch East Indies and the founder of Batavia (now Jakarta), justified the Dutch exploitation of native populations this way: 'May not a man in Europe do what he likes with his cattle? Even so does the master do with his men'.[15] In this way, cheap labour transformed cheap nature into a ready and abundant source of wealth.

Looking along the timeline of colonialism, we see a transition from cheap nature to cheap labour to, now . . . cheap data.

Cheap data involves the same extractive rationality that we saw with cheap nature and cheap labour. Think of how much data we generate through our activities online, uploading and tagging content, liking posts, sharing things, posting our own content. All of this takes our time and effort, sometimes too much of it. But in the hands of the platforms, it arrives absolutely for free, without payment. Such data is said to be abundant, without owner, and 'just there' for the taking. Yes, you and we might produce it individually, but by the time it is aggregated, it is considered an owner-less exhaust or by-product, over which we can no longer exercise any claims.

Like many other colonial resources, however, data needs processing, refining. That job is technology-intensive and requires highly specialised tools. Which means only big and sophisticated corporations can do it. Our only job, it seems, is to generate the resource. We are told that this is what progress looks like, that our data is used for connecting us better, building better communities and fuelling the AI that will solve humanity's problems. But so far the real beneficiaries are not us; they are the owners of the platforms and AI models that help maximise their profits.

Data and the Continuation of Colonial Violence by Other Means

We said that the colonial data grab of the twenty-first-century economy can be effective without violence, and explained why that is. But that does not mean there is no violence associated with data colonialism. Far from it.

Yes, data colonialism exhibits different forms of violence from those encountered in the past. As we know, the inhabitants of the colonised lands were discarded or eliminated by the millions. But instead of the vicious physical brutality of historic colonialism, data colonialism makes possible novel forms of *symbolic* violence,[16] such as those that come with discrimination, loss of opportunity and the classification of people into disadvantageous categories by AI and its algorithms. These forms of symbolic violence can mesh together to have long-term effects that are indeed physical. Being passed over for a job or being denied certain welfare benefits by an automated decision system, while initially a symbolic act, can have very concrete and physical consequences for people's health and wellbeing.

In other words, data colonialism has its own distinctive forms of oppression. Violence in historical colonialism was organised through racial hierarchies, and its impacts were felt through brute force. Although violence in data colonialism may be less brutal, and its particular tools and techniques may be new, its impacts are still concentrated in certain populations, and they often continue the legacy of racist, sexist and class violence that started with historical colonialism. Vulnerable populations within these categories are, after all, the most likely to be subjected to more pervasive forms of surveillance at work or school, or rely on precarious forms of gig employment, or be subjected to algorithmic-promoted hate speech because of their identity or status.

Nor is algorithmic discrimination the only type of symbolic violence associated with data colonialism. As a social order, data colonialism also attacks and undermines alternative forms of thinking and of being. This is a complex point that will gradually become clear as the book's argument unfolds. But our basic claim is that, as human habits become colonised by Big Tech and digital platforms, it becomes easy to forget that there once were other ways of being social before extractivist platforms came along; to forget why it is we feel compelled to install apps on our phones, or why we don't question who is collecting the data generated by our smart cars or appliances. After all, everybody else is doing it! Nobody wants to feel left out, or be what evangelists of Big Tech call a miserablist (someone who only sees the negative side of these new technologies). But if we lose the habit of questioning the assumed benefits of these products, we risk forgetting what the point of resisting the new order would be.

What then are the specific harms that data colonialism commits?

Why the Effects of Data Colonialism Will Never Be Evenly Distributed

Some people have more means to protect their privacy. Maybe they can afford an Apple device, which is more expensive but which harvests less data than other brands of devices (or at least keeps that data within the Apple corporation, instead of selling it to a third party). Or maybe they know how to install ad blockers or VPN services. Or maybe they are just the type of wealthier consumer who tends to win out when ad-targeting produces good prices for them, or when algorithms make decisions that give them exactly the services they want ... to put it crudely, the ones who employ gig workers, not gig workers themselves.

Needless to say, this inequality is nothing new. Historical colonialism wasn't just about making vast amounts of money, but about reinforcing which types of people had access to the world's resources and which didn't. The same impulses are exhibited in data colonialism. There is, and will continue to be, a data colonial elite, and its composition will be heavily shaped by the legacy of the historical colonial elite.

But even for those who do feel they benefit overall from our deal with Big Tech, it is important still to acknowledge that platforms and algorithms are not as kind to everybody else. But who is this 'everybody else'? Who does not benefit to the same extent as we do when participating in data colonialism? For the foreseeable future, they will include the people who have always paid a heavier price during colonialism: the poor, people of colour, and women. Colonialism may keep evolving, but its unequal legacies and injustices go on being perpetuated.

This continuity makes even more sense when we remember another key feature of colonialism: social classification. Classification is a general feature of society, and doing it in a new way through automated decision-making potentially affects everyone. Since data is exactly a tool for discriminating between things and people, it is no surprise that bias and prejudice are replicated when data is collected and fed to a system that can make automated decisions about the welfare of already vulnerable groups. Indeed, this kind of systematic discrimination, which allows one group of people to classify and control the rest, is very much a colonial invention. Strategies of divide and conquer by colonisers have for centuries been partitioning societies along the lines of class, gender and race: rich above poor, men above women and gender non-conforming individuals, and white over everybody else (and also everybody else against each other, since hierarchies promote competition). This structure was and still is based around a figure at its apex: the superior rational authority of the (white, male) coloniser. The racial segregation behind Jim Crow laws in the US, behind apartheid in South Africa, or behind the British Empire's colour bar (perfected in the colonies and then imported back to the centre of the empire), are all examples of this kind of hierarchy in which rich white men are at the top.

Class, gender and race have thus always been instrumental in creating the myth that human differences need to be managed and ruled by a universal Western rationality. And data in some form has for a long time been a part of this project, as it provides the means to quantify and manage difference: census data, surveillance data, demographic data.[17] In this respect, once again, data colonialism renews the mission of historical colonialism through fresh means – algorithmic decision-making, digital platform design – that are even more subtle and harder to trace than before. Data, after all, always discriminates. You can't build a database of Xs, Ys and Zs, unless you have already decided that this is an X, and not a Y or a Z.[18]

The Colonial Continuities of Data Discrimination

Algorithmic discrimination based on data extraction is often used against those struggling with poverty. This is already happening in rich countries, and it is likely to spread more widely as algorithmic decision-making and its techniques are increasingly imposed by corporations and governments. In her 2018 book, *Automating Inequality*, Virginia Eubanks analyses public assistance systems related to Medicaid, homelessness, food assistance and assistance for at-risk children in the United States. In many of these systems, who gets access to the services is determined automatically by algorithms, whose operation is not transparent and seemingly arbitrary. Eubanks describes the punitive and harmful effects inflicted on those who are deemed ineligible for these benefits. Class intersects here with race and gender, which brings to light the demeaning and excessive amount of data surveillance that is inflicted on the poor (and more pronouncedly on poor Black women).

As a result, it is those at the bottom of the social pile with only a cheap phone and limited contract, who depend on that phone and its apps to get their next insecure job, who have literally no chance of renegotiating in any way the privacy-constraining conditions those apps and their employers impose upon them. All this may happen even against the best intentions of those devising the data systems, because of the wider social power inequalities in the midst of which they are used, and the huge information asymmetry built into the platforms themselves.[19]

This is also painfully clear when we turn to gender, where the violence of extractivist platforms is just as harmful. The effects range from the destructive influence social media algorithms can have on teenage girls and their sense of their own bodies (an influence platforms like Facebook seem to be fully aware of, but which, according to the testimony of whistle-blower Frances Haugen to US Congress in October 2021,[20] they refused to regulate in order to maximise profits), to the increased 'dystopian' workplace surveillance that women and minority workers endure,[21] to the general harassment and bullying that women experience online because of platforms' failure to put basic protections in place.

Take pornography. The connection with data colonialism might not be immediately apparent, but consider how datafication has allowed highly profitable companies – Pornhub received around 4.5 billion visits per month in 2020[22] – to prioritise the monetisation of content over the welfare of women. There is a standard practice on those sorts of platforms of allowing for the posting of abusive content (sometimes involving minors) anonymously, without the subject's consent, and without effective recourse for having it removed. There is also the predatory practice of deepfake porn, using

AI to realistically superimpose women's faces onto pornographic material without their consent.

But digital violence against women goes well beyond the non-consensual dissemination of sexual content. Writing from Mexico, a country where on average ten women are murdered every day, Grecia Macías identifies the multiple ways in which women experience various forms of digital violence, including 'unauthorized access to one's devices, digital threats, smear campaigns, extortion, attacks on one's means of expression, identity theft, the manipulation of information, surveillance and stalking.'[23] Data from around the world provides plenty of examples: according to a study involving young women in Ghana, Kenya and Vietnam, 70 per cent of them report being exposed to censorship, verbal abuse, stalking, extortion and violence when using social media to seek health-related support or sharing health information;[24] 73 per cent of women journalists have experienced online violence; half of young women in the UK have received unsolicited sexual images;[25] and non-white women candidates in the US are twice as likely to be the targets of mis- or disinformation and four times more likely to be the victims of online abuse than white candidates.[26] It is true that most of this violence is perpetrated by individual males (although when it comes to surveillance and censorship, the state is also a major aggressor). But as with the example of non-consensual sexual content, platforms definitely play a role in minimising or ignoring the risk that their technologies and business models represent for women (in a recent example, two women sued Apple, claiming that AirTags made it easier for stalkers to harass them).[27]

These impacts are often further compounded when it comes to gender non-conforming people. In addition to the forms of violence mentioned above, they may be harmed by the very idea on which data-extracting corporations build, which is to fix people into this category rather than another. This process is what the information scholar Anna Lauren Hoffmann calls 'data violence', and it is

automatically meted out to those who don't accept being catego-
rised in simple gender terms.[28]

For all of these reasons, it is important that, as Catherine
D'Ignazio and Lauren F. Klein argue, the design, production and
evaluation of digital technologies is informed by a data science
aligned with feminist values and practices, which can help to both
expose inequalities and change the distribution of power.[29]

Race, as has recently become very clear, is also a crucial factor
in determining who benefits, and who doesn't, from data colonialism. There are many examples of how people of colour (or POC)
undergo algorithmic discrimination in everything from insurance
rates to targeted advertising to court sentencing to police surveillance, and more. Some effects are more indirect. For instance, a
recent ProPublica report uncovered an algorithm helping landlords
in the US set the highest possible rents on tenants, exacerbating the
current housing crisis in a way that is likely to impact tenants differently depending on their race.[30]

Consider another algorithm used widely by US hospitals and
insurance companies to determine the level of healthcare that
patients were assigned. In a 2019 study published in the journal
Science,[31] researchers found that the algorithm, used to manage the
care of over 200 million people, exhibited a racial bias. Apparently,
when considering patients with similar ailments, the algorithm
more frequently referred white patients than Black patients for additional healthcare. In short, Black people had to be sicker to receive
the same healthcare than their white counterparts.

How could this happen? The algorithm assigned a risk score to
a patient based on the amount of money the insurance company
spent on them, and then recommended additional care based on
that risk score. But because of the powerful background of institutional racism and Black people's resulting distrust towards the
healthcare system, they consume less health resources than white
people, which meant that simply relying on such data guaranteed

them a lower risk score even when they were, in fact, more sick than their white counterparts. The result was lower recommendations to Black people for additional care.

Are racist engineers designing these systems intentionally to behave in racist ways? Not necessarily. We have to remember that these systems don't operate in a vacuum. In order for an AI system to 'learn', it must be trained with lots of examples from the real world. And if those examples reflect the bias and discrimination that we find in our real world, guess what's going to happen. Predictions based on those examples will reproduce that underlying discrimination. Differences in outcomes from data-driven systems can therefore be not just a function of preference or intent, but of deeper social structures like racism.

The result, indisputably, is that data systems of all sorts reinforce stereotypes and institutionalised discrimination. Indeed, data scientists themselves have proved that narrow approaches to reforming AI Ethics miss the wider structural forces that shape the inputs to algorithmic decisions; as a result, we can't rely on algorithms themselves to generate justice.[32] Furthermore, the fact that the design and deployment of these very complex systems are generally in the hands of small numbers of privileged white people (itself the long-term legacy of colonialism among other things) means that the voices of the rest of society tend not to be heard, at least until the damage has already been done.

That most of us do not even think of questioning the premises under which class, gender and racial discrimination operate in data systems is testament to the fact that inherited colonial ways of viewing the world are firmly entrenched in our thinking. In addition, our unwavering belief in the rightness of data extraction and Big Tech's supposedly superior rationality leads us to trust whatever civilising mission the computer scientists put forward to 'improve' our lives, overriding our doubts.

It is not that data itself is bad – how could it be? We need data and information about disease, about the stars, about the flora and fauna that surround us, about the damage we collectively do to the planet. But the issue is *how* data is extracted from what, from whom, and on what terms.

It makes no sense to be against the idea of data in the abstract. But a social order for extracting data from human life under very particular conditions that benefit the few and not the many is another matter entirely: it is data colonialism.

The Colonial Roots of AI

The social structures that today make the continuous extraction of data from our lives a comfortable and even natural reality have been in development for hundreds of years. The business models and scientific principles endorsed by Big Tech and AI seem to have emerged quickly and out of nowhere, but they are firmly rooted in processes that are much older and well established. As we began to see in the earlier discussion of cheap data, colonial impulses do not simply disappear one day: they get reconfigured and refocussed on new processes.

To start tracing these continuities, we need to recognise that Western science, as well as capitalism, co-evolved side by side with colonialism. This is a subject that has filled volumes, so here we will simply point out that the need to manage colonies during colonialism was satisfied by the development of forms of knowledge that made the management of those colonies more efficient. And these forms of knowledge – that is, modern Western sciences such as botany, geology, anthropology and zoology – resulted in technological breakthroughs that allowed corporations to rule the world. Let's take a closer look at how this process unfolded.

According to Steven Harris, the operation of long-distance corporations in colonialism required substantial investments in science.[33] Enterprises such as the Casa de la Contratación de las

Indias or the East India Company were tasked with managing colonies remotely, across great distances. Mexico had to be managed all the way from Madrid, and Delhi from London, at a time when the handwritten or printed word was still the main means of communication. So new tools had to be developed by scientists and inventors employed by these corporations to make the long-distance management of colonies more efficient.

While the standard view of history encourages us to think of science as the outcome of contributions by individual male geniuses (Galileo, Bacon, Newton, etc.), Harris shows that colonialism itself helped to introduce a kind of Big Science developed by a collection of lesser-known scientists. Those contributions had just as much impact as that of the scientific celebrities, but they get a lot less attention. A global network of scientists and practitioners worked in concert for the benefit of their employers or institutions. Much of this went on in the colonies. For example, between 1717 and 1738 Spain imported 45 tons a year of medicinal plants from the Americas, a number that increased to 155 tons a year between 1747 and 1778. The knowledge of how to use these plants came from indigenous peoples, but was systematised and commercialised by apothecary missionaries in the colonies, as the name of the popular fever-controlling plant of the time, cinchona, known as Jesuits' bark in Europe, implies.[34]

By the end of the seventeenth century, these Big Sciences (medicinal botany, observational astronomy, geography, natural history, meteorology, navigation, etc.) had 'modernised' themselves into what today we would call distributed or virtual teams. There were Royal Societies or Academies for this or that discipline, which invented new technologies and tools, sometimes adapted from or inspired by pre-Columbian inventions in medicine, architecture, astronomy, transportation, and so on (from cable suspension bridges to syringes to oral contraceptives).[35] They also generated vast amounts of data stored in reports, charts, maps and handbooks, and curated great repositories of specimens like the botanical gardens of Amsterdam and Leiden, or the Royal Botanic Gardens at Kew in London (which developed new strains of coffee, cocoa, maize and various fruits that were better suited to local diets).[36]

In other words, colonial expansion and administration was carried out by global corporations (a sort of Big Tech) employing hundreds of scientists and technicians (Big Science) who developed the intellectual and technological structures required for the political, military and cultural subjugation of the world, in the interest of profit making. This required the collection and analysis of vast amounts of information from the colonies (Big Data), which became an important asset for their management.

Much of the information that was collected by the colonisers was applied in the surveillance and control of colonised peoples (Big Brother). This extended into the political management of populations too, through mechanisms like the construction of security fences to divide populations, the panopticon design of jails to place prisoners under constant observation, and various sorts of early forms of biometric data collection – all of which became commonplace in the colonies before being implemented in European cities.[37]

It is therefore the long history of colonialism, not just the last few decades of the Digital Revolution, that is the best starting point for understanding how these processes are converging. Already, centuries ago – long before GAFA (Google, Apple, Facebook and Amazon), their Chinese equivalents BATX (Baidu, Alibaba, Tencent and Xiaomi) and the rest of the players that make up today's data colonial class – the cumulative inventions of Big Science throughout modernity allowed Big Tech to use Big Data to act as Big Brother. Which means that what's going on with data is not just capitalist business as usual, or even a problematic recent variant of capitalism which needs taming. Large-scale data extraction, whether to run platforms or fuel generative AI, represents the continuation of a long-established way of organising the world's economies, societies and power relations on a profoundly unequal and exclusionary basis.[38] Within this longer context, we can see clearly that the extraordinary scale and impact of today's new data extraction has, in fact, a clear parallel in history: the start of historical colonialism.

The Resilience of Colonialism

This historical framing helps us see beyond the scandals and market uncertainties that Big Tech periodically confronts.

Take the recent claim that Big Tech's 'free ride' might be coming to an end. Although Amazon, Apple, Facebook, Google and Microsoft posted record profits during the pandemic – US $1.4 trillion in 2021, or a 55 per cent increase from a base that was already quite high[39] – recent trends seem to suggest a slowdown, as well as a host of other problems for Big Tech.

This claim seems to be reinforced by evidence that data privacy rules are being regulated more effectively across the globe, curtailing extractivism to some extent. Europe's GDPR is a good example; in addition, 35 out of 50 states in the US are at least considering some form of privacy regulations.[40] Indeed, the stronghold that Meta (Facebook) and Alphabet (Google) have on the digital ads market seems to be diminishing, decreasing from a 54.7 per cent share of the market in 2017 to a predicted 43.9 per cent by 2024.[41] Still quite substantial, but a decrease nonetheless, and perhaps a sign of healthy competition. Furthermore, legal woes are starting to catch up with companies like Meta, who recently settled its Cambridge Analytica scandal for US $725 million, but then suffered European rulings that challenged how it exploits user data for advertising.[42] All major tech companies have pending lawsuits throughout the world in areas such as competition, privacy, misinformation and disinformation, illegal and abusive conduct including sexual exploitation, and algorithmic discrimination.[43]

For start-ups, the situation has also recently become difficult, as venture capital dries up and forces them to enter into less favourable debt deals.[44] All in all, some experts are reconsidering their ideas about seemingly endless high growth rates in this sector (although many of these companies remain extremely profitable).[45]

Does this mean Big Tech will wither and disappear, and data colonialism will collapse before it has even reached its full scale? Not likely. One of their responses to this crisis has been simply to fire workers. US-based tech companies laid off 131,000 workers in 2023 (as of March, with more expected in the rest of the year) and more than 93,000 in 2022,[46] although many of those sacked by Big Tech companies are finding employment within the general economy.[47] After Elon Musk acquired Twitter, 50 per cent of its workforce was laid off.[48] This is a good reminder that as fortunes expand and decrease, it is workers, not shareholders, who bear the brunt of hardship. One need only look at the poverty-stricken slums that emerge around Amazon warehouses in places like Tijuana, Mexico, to get a visual reminder of this fact: next to state-of-the-art Amazon facilities stand shacks built out of tin, cardboard and plastic, where workers live. A similar scene is replicated across communities where Amazon operates, although salaries and living conditions might be a bit better in the Global North. Amazon receives tax breaks and other incentives to construct their warehouses, on the premise that they generate new jobs. But the jobs that Amazon creates are poorly paid (with hourly rates of around US $2.60 in Tijuana), and with less benefits for workers than the jobs they frequently eliminate.[49]

Another response to crisis might be simply to double-down, reinforcing current extractivist practices so that consumers and citizens, not just workers, pay the price. This is what Naomi Klein called 'disaster capitalism': the idea that corporations and governments take advantage of periods of natural or financial disaster to impose more oppressive measures.[50] The Covid pandemic is a good example of this dynamic. According to research conducted by the European Center for Not-for-Profit Law, the International Network of Civil Liberties Organizations and Privacy International, surveillance technologies were misused by governments during Covid emergency periods by enabling the abuse of personal data, the silencing of dissent, the repurposing of counter-terrorism measures to track civilians and the extension of these measures beyond the pandemic.[51] These four trends were accompanied by a fifth one: the increasing influential role of the private sector as a provider of these extractivist technologies.[52]

The lesson is that, whatever economic slumps the tech sector might experience along the way, they – in partnership with governments – will probably find ways to externalise the threat and pass it on to the public, often in the form of amplified data extractivism practices that affect workers and consumers. That is why the current economic downturn probably won't spell the end of data colonialism. Nor will the collapse of large social media platforms like Meta, or even the end of surveillance capitalism that some people predict (we ourselves are sceptical),[53] mean the end of data colonialism, because, as we will show, data colonialism is much wider than either of them.

We Need Not Be Passive Victims

This is only the first chapter of the book, but the problems posed by data colonialism may already seem insurmountable. What to do but throw up our hands in horror and then go on just as before? In any case, isn't our data already out there in the hands of corporations?

We get that reaction a lot. After all, if these are the same problems humanity has failed to resolve in 500 years of colonialism, what are the chances we will succeed in resisting colonialism's new formation? We will continue to detail the serious, sometimes devastating, impact of data colonialism over the next few chapters, so there is still a lot more bad news to cover, unfortunately. But we also want to give a sense of what the starting points for resisting this emerging social order might be.

Even at the beginning of historical colonialism, there were people who questioned it and resisted it. Colonial subjects have always had this power, a power that in the end led to overthrowing the political structures of colonialism, and so do we. The coloniser simply cannot exist without the colonised, so the latter have always had the power of refusal (even if it has come at a very high cost, including often the ultimate sacrifice of life itself). To begin exploring this resistive power, let's consider our own role in this system.

As colonialism developed, a Eurocentric worldview became enormously influential, shaping the consciousness of the colonised and pushing them to believe that Western power was incontestable, that Western science was the only form of reasoning capable of explaining the world, that Western religion was the only one capable of saving the soul, and that Western culture was the most advanced expression of humanity. Colonialism, in other words, gave the colonised a sense of inferiority that has taken – and continues to take – a lot of work and time to overcome.

Similarly, when it comes to the new data colonialism, it is easy to feel that we don't have the freedom or opportunity to do anything but accept this new world order; that we have no choice but to click a button that says 'I accept'. But while there is a lot of deception and exploitation at work in getting us to click that button, and some of us need to click that button just to earn a living, the fact is that we are not helpless. Put simply, if we accept something, we can also un-accept it.

And that's for a clear reason. In the end, data colonialism needs us. Without our data, it does not work. Most data is not extracted without our consent in some form, and we are thus participants in the act of extraction. True, stopping our consent may only help us against future extraction, not the impacts of all the data previously extracted that affects us, but that does not mean Big Tech no longer needs our consent. Big Tech works by us not fully understanding the implication of being complicit in its extractivist systems, but for now we do remain complicit to various degrees, depending on our situation. In order for anything to change, it is important that we first recognise the full extent of our complicity in what is unfolding.

When reviewing the history of colonialism, some talk about the role of the 'native informant'. These were members of the colonised class who recognised early on who would be on the 'winning' side of history, and decided to align with them, acting as translators not just linguistically, but culturally. A good example is La Malinche (aka Malintzin, aka Doña Marina), the Nahua slave woman said to be gifted to conquistador Hernán Cortés who acted as his translator, consort and advisor during the conquest of Mexico. While it is easy to portray such informants as traitors, we cannot guess what their actual conditions and intentions were. They were probably just trying to survive the collapse of their world, and the emergence of a strange new one. In cultural lore (at least in Latin America), these informants are seen as the native mothers who, along with the European fathers, gave birth to a new hybrid race of people (not necessarily out of love, but out of violence).

But in data colonialism *we are all informants*. We literally translate our lives for the coloniser by conducting our social lives on their platforms. This is the disturbing realisation to which Tracy came after some reflection. However, that's not the whole story. One advantage we have at this point in history is that we already know what the colonial system wants, and how it works. We have lots of examples of how colonialism can be resisted, and we can still resist if we are clear about what is really going on.

But first, let's take a closer look at how this dispossession is unfolding in our daily lives.

CHAPTER 2

DATA TERRITORIES

NOW IN HIS forties, Michael had promised himself and his family to lose some weight and live a healthier life. So, when his employer offered to enrol him in a wellness programme and give him a free digital fitness tracker, he didn't think twice about it. The device looked cool and sleek on his wrist, and he had received many positive comments about it from his friends. People couldn't believe it was free. His doctor was pleased, and mentioned it would be a great opportunity to track those blood pressure issues to which, as a Black man, he should be paying more attention. Bring on the information revolution, he thought.

He soon found himself checking the device every few minutes to measure his exercise activities, sleeping habits, caloric burn rate, blood pressure, and more. He would get a bit annoyed when the device admonished him for not meeting his daily activity quotas. Work quotas, exercise quotas, quotas all the time! Although maybe that was a good thing, he thought. Helps keep me on track.

But Michael also started noticing some weird things. Why did he start seeing online ads for the healthy smoothie place he passed during his walks? Also, his employer announced a new initiative at work to combat stress, which data had shown was on the rise at his particular office branch. What data? Were they using data from the fitness monitors?[1] Another time, after a night of heavy drinking (which he regretted), he noticed the device showed he had been burning a lot of calories. 'Oh yeah,' said his co-worker Fred, 'I noticed it does that, probably because drinking elevates your heart rate. So, it probably knows when you've had one too many.' That's a bit creepy, thought Michael. He also read in one of his Facebook groups that the particular fitness monitor he was using was 'racist', meaning that the device (which uses a green light to measure blood flow) was not able to generate accurate readings in users who have dark skin, a design flaw that the company had not shown much interest in addressing.

But that wasn't the half of it. Somewhere in the terms of use, which he hadn't really read, Michael had given permission for *all* of his health data – in fact all of his 'health-*related*' data – to be shared. This meant that, potentially, data collected by the tracker that had nothing to do with Michael's health (such as his location or activities) could be shared not only with his employer, but with marketing firms, the authorities, or even stolen by hackers.[2]

As Michael's story illustrates, it's not exactly news that there's something problematic with how data is collected and used. We often first hear of this via stories of how our data is not safe from hackers. Think of the data leaks through poorly secured data storage: in 2017, a data breach at leading US credit broker Equifax exposed the records of 147 million US citizens and millions of others in the UK and Canada, which was just one of many data breaches across many countries in the past decade.[3] But we also learn about the problems of data mining through stories about poorly designed algorithms that silently discriminate against us. Even mainstream business outlets like the *Financial Times* talk critically of the 'surveillance model' of Big Tech businesses: making money by tracking us and our data.

Are these just the teething troubles that any major new industry can expect to face? Or perhaps the exploits of some rogue players that need reining in? Isn't the right way of handling this to avoid putting anything personal on Facebook, switch from an Android to an Apple phone (if you can afford one), and trust in the government to bring in some reasonable privacy controls on Big Tech?

The answer is no. Such piecemeal measures miss the real point, which is that there is a much wider landgrab of data going on across societies today.

Historical colonialism was nothing without the control of new territory. It was this territorial control that made possible the extraction of new resources on a global scale and the creation of vast new global markets for the benefit of the colonisers. The regions we now call South and Central America, the Caribbean, and North America seemed like a 'New World' to their European colonisers. But they were no newer than Europe was, and they were inhabited by First Nation peoples who generally had to be displaced for colonial processes of extraction to begin. There were multiple models for generating the labour needed to exploit those new territories at scale: the importing of slave labour to the plantations of the Caribbean and parts of South and North America or, alternatively, the sending of settlers from Europe to live on the colonised land. With Britain's capture of India, a further colonial model developed: absolute control of the terms of trade with India and the managing of the local economy through military force. A combination of these different models spread to Asia, Africa and the Pacific. But whatever the model of labour relied upon, historical colonialism depended on the control of territory.

The volume of land acquired and controlled was extraordinary: at the height of their colonial power, Spain controlled 5.3 million square miles beyond its national borders and Great Britain 13.7 million square miles.[4] The economic impact of the additional land that European countries were thus able to exploit was huge, although estimates vary: one scholar argues that it increased *sixfold* the land per capita of the European colonising nations.[5] This territorial expansion was the basis not only for economic exploitation but for completely transforming the everyday life of the controlled territories.

But data is not like land: you can't touch it and you can't walk around in it. It might therefore seem strange to talk about *grabbing* data in the way that colonisers once grabbed physical territory. Indeed, data can be endlessly copied and reused; it is in principle what economists call a 'non-rival good', so multiple parties can use it at the same time, making it seem even more remote from physical things like land or bodies.

So, if data is today's new type of colonial resource, how can a 'territory' be created that is controllable in ways that ensure data's maximum extraction? In short, by writing code. More precisely, by writing the specific code that builds new computer-accessed spaces and forces any interaction within those spaces to unfold in ways that enable the continuous extraction of data about those interactions. We will call those spaces *data territories*.[6]

In everyday life, we know those data territories generally as digital 'platforms' – think Facebook and Amazon. But as we will see, there are many other types of data territories. The interface of a search engine like Google is another type of data territory, as are the interfaces of countless 'smart' tools and devices. All of them give us access to a commercially controlled space from which data can be exclusively and continuously extracted from us. We are not just talking about single territories, but about networks of data territories: through software, and without any physical constraints, any data territory can be connected to countless others to create a vast archipelago of connected data territories. Multiply those connections still further and the whole surface of social life becomes covered many times over by relations of data extraction.

Even though data is not land, data territories can be controlled by power with a completeness that rivals, perhaps surpasses, that of the original colonial rulers. The seizure of data territories therefore becomes the start of a larger transformation, both economic and social, that builds on these new relations. Just as colonial territories imposed new laws on their inhabitants, so data territories exert their own 'laws' based on the code that keeps the platform running.

When Society *Becomes* the Territory

How did data colonialism's *new* territorial capture begin? There was nothing quite as dramatic as the physical discovery of supposedly new continents. Data territories have their basis in software that is embedded deep within the devices we use and the non-physical 'spaces' (the apps or platforms or websites) that we access through those devices.

But the emergence of data territories was rapid and dramatic nonetheless. Together, they have opened a 'New World' of data extraction that offers not just data, but a completely new way of organising the relations that make up society, quite literally reorganising society for business. How? By directing our activities into social channels that are fully under businesses' control. Data might be a non-rival good, but what it captures – human life – is definitely a finite resource, which, at root, is what makes this an exploitative process.

We, as consumers of digital services, certainly don't know much about the data-processing operations that go on behind the scenes in a data territory. But if we try to leave that territory, or do things within it that don't conform to the territory's rules, all sorts of legal and practical constraints get put in our way. In that sense the data territory is definitely bounded like a physical territory.

The steps through which such data territories came about map directly onto the first three steps of the Four-X model mentioned in the Introduction: Explore, Exploit, Expand. Signs of the fourth, Exterminate, are also emerging, as we'll see later in the chapter. Let's continue our application of the Four-X model to connect the past of historical colonialism with the present of data colonialism.

Exploring New Worlds of Data Capture

Just as the original European explorers didn't know they were about to discover wholly new reserves to be exploited (Columbus believed he would find new routes to the already-known India), so too the territories of data colonialism emerged by an indirect route.

It all started with a very basic and banal feature of computers: they capture data. To explain how computers capture data, we must look back over the internet history of the past 30 years from the point of view of social transformation, not just technical evolution. Computers work by capturing organised sets of information and storing them, in dynamic and adjustable form, such as the document that we produced by typing this chapter into our laptops.

Seems innocuous enough. But the US information and privacy scientist Philip Agre, writing in 1994, saw the deep social significance of this new type of information tool. For computers do not just capture information in the abstract: they capture it *on their own terms*. Unless I interact with my computer in very particular ways that comply with its specific rules, nothing gets done. Smile at the keyboard, and nothing happens; I have to move my fingers over the keyboard in particular ways to produce inputs that comply with sets of computer-based rules that Agre called the computer's 'grammar of action'.[7] This means, in effect, that computers always impose their grammar of action – their particular template for how you interact with them – on whoever uses them, and that grammar always includes the capture of data. This is the root of the often-repeated idea that 'code is law':[8] platform code, in effect, sets the laws of data territories.

In fact, a computer can capture traces of any interaction, provided the interaction takes a form that fits with the computer's grammar of action, that is, it is encoded in accordance with the computer's rules of use. So, if my laptop has a camera embedded in it and the camera is switched on, then my smile can be recorded and stored. But Agre had in mind much more than the capture of what individual people do on their own computers. He was writing at a time when computers were beginning to be connected ever more easily, in what we now know as the internet. The World Wide Web's design had also recently been announced, which would make particular types of captured information placed at particular computer-based locations (websites) searchable by other computers.

What Agre sensed right at the start of our digital era was that, if all computers become connected, then any computer could, in principle, track the contents captured by any other computer. The consequences of this for everyday life would only grow as people started to use computers more and more across daily life, and those computers began, all of them, to be connected to each other via the internet.

The year of Agre's article, 1994, was the first full year of operations of Netscape, the first commercial browser. It was also the year when Lou Montulli, an employee at Netscape, invented the 'cookie'. The cookie was initially a device to save us having to keep inputting the same data to a particular website every time we visit it: it certainly is a nuisance when a retail website doesn't remember your address and phone number from last time you visited. But advertisers soon realised the huge potential for data capture of a technique whereby your clicking on a website generates a cookie sent to your computer that then stores certain information (on your computer, but accessible to a distant computer associated with the website). This was the start of all the tracking mechanisms that marketers placed on the web and on our phones that have recently caused a backlash from Apple and others.

In the early 1990s, when Agre was writing, no one yet knew the intensely commercial path that the internet and World Wide Web would take. But already by the late 1990s, a completely new type of online society was coming into view that would be dominated by commercial corporations and their imperatives for profit. To achieve this, all that was needed was for individuals to live much of their lives on, and through, connected computers, and for businesses to take advantage of the traces stored on those individuals' computers. Two things now so banal that we can't imagine life arranged any other way, but here we have the origin of today's data territories. Data colonialism's phase of exploration had begun.

Exploiting Data Territories

Far-sighted legal and communications experts were already warning in the late 1990s that the long-term result of allowing private corporations to manage computer connections on a large scale would be a new structure of power with negative implications for democracy and basic values like privacy. Why? Because a networked computer space dominated by commercial imperatives would set no limit on how computers can track each other's information for commercial advantage; it would set no limit on human beings' vulnerability to continuous data extraction.[9] Those experts sensed that the whole of everyday space could be transformed into what, looking back, we can see as data territories.

The story of how this possibility became real has been told many times. You will have lived parts of that story yourself. It took a lot of machinery. One piece of machinery was the Google search algorithm, which quickly became not just the world's dominant search engine, but also the source of personal data that fuelled Google's enormous advertising business. Google had realised that the information about us it had accumulated in order to serve us with more relevant search results was potentially of huge value in predicting other things about us, not just what we searched for. But behind this and similar machinery lay a completely new type of economic exploitation.

Few described this opportunity as clearly as Google's own chief economist, Hal Varian. Google's data territory is by no means the only one that matters, but we mention Varian because, in a paper for a dry academic conference, he laid bare what we can now see as the exploitative inner logic of the data territory: how it enables corporations to use data to exclusively track and influence their customers. This territory (even the very idea of it) simply had not existed until two decades or so earlier.

Varian's description started with something very simple: there was 'now a computer in the middle of virtually every transaction'.[10] A *connected* computer, of course, since a computer without an internet connection was of no use to Google. What followed were certain new social processes that Varian thought important:

- Data extraction and analysis
- Personalisation and customisation
- Continuous experiments
- New kinds of contracts due to better monitoring of what customers do[11]

Data extraction is obvious: that's the basic thing that, as we've seen, connected computers can do. Personalisation is also familiar: ad companies and social media platforms constantly tell us that by gathering more data about us, they can make their services more personally relevant to us.

But what about experiments? By 'experiments' Varian meant the possibility that platforms could use their control over computer-based transactions to systematically try out what pricing or other signals would be most effective. As to new contracts, Varian had spotted that, because every transaction conducted online was now directly observable, information streams that were before out of business's reach suddenly became direct inputs to business. This changed the power relations between businesses and customers profoundly, because corporations could dictate interactions on their own terms (such as signing away some of your rights as an individual just by using a platform) and, in that way, secure your presence in their territory, a territory for continuous data extraction.

Put all this together, and you had a completely new space of commercial power right at the heart of everyday life. A new type of power relations had emerged based on the extraction of data.

Yet Varian was simply setting out the implications of the business model that his own company had long since pursued. Varian was one of those behind the design of Google's system for auctioning ads, in real time, by drawing on the precise data Google possessed about a user's likelihood to click on that specific ad (they had, after all, consistently been tracking what you searched for). Varian saw that every commercial transaction would over time come to operate on the basis of this new sort of tracking-based corporate power, that is, via exploiting data territory.

Varian also saw something else: that this new model of economic exploitation would have the potential to sweep away other forms of economic interaction that extracted less data. Here we see the seed of data colonialism's fourth stage (as per the Four-X model): Exterminate. We'll return to Exterminate as we unfold the signs of how data colonialism is starting to override older models of economic and social interaction. For now, just consider Varian's seemingly benign example: the traditional insurance contract. What happens, he asks, when user behaviour can be continuously tracked by insurers?

> When I rent a car, somewhere in the fine print there is a statement to the effect that I will operate the car in a safe manner. But how can they verify that? There used to be no way, but now insurance companies can put vehicular monitoring systems in the car . . . They can use these systems to verify whether or not you are fulfilling your part of the contract. They get a lower accident rate and I get lower prices.[12]

Sounds rather like institutionalised surveillance, and indeed that's what it is. And it has become the prevailing model today: 'usage-based' insurance is now booming, especially vehicle insurance. Yet Varian ignored the obvious consequence that the connectedness of the internet propels all of us into a space of continuous surveillance – not by the state, but by corporations – a space where all sorts of data about us can be used by insurers in new ways in data territories of insurers' own making.

Shouldn't safer drivers enjoy lower premiums that reflect their driving more closely than through the old-style no-claims bonus? Sure, but every driver makes errors and has bad days. Such performance drops may have very different consequences in a data territory: whereas once you would have just berated yourself and hoped no one noticed, if you have a smart insurance policy, you can be sure your car noticed and stored the data for the benefit of the insurer. How this will impact the evaluation of your driving will remain opaque to you.

The Expanding Data Colony

Data's modes of expansion are technical and software-based; they relentlessly expand the scale of data territories, in the process completely reorganising the spaces where we live and work. Three techniques have proved crucial.

The first was to expand the range of computers we use in daily interactions. Remember Varian's point that 'there is a computer in every transaction'. Well, in the opening decades of the twenty-first century, this became ever more the case, as more and more people carried small computers with them that could connect online: their phones. Since then, tablets have emerged as more easily portable versions of laptop and desktop personal computers. Ubiquitous computing has moved from being the vision of specialised computer scientists to being an everyday reality, in richer countries at least.

The second technique was to create platforms: special-purpose spaces of computer-based interaction where pre-determined types of transactions and interactions can go on, while data is continuously extracted from people. That's what Facebook is; that's what every app is. Platforms were already thriving when Varian wrote his paper, but they were only just beginning when Google made its first moves towards monetising its own search data in the early years of the century. Platforms reconfigure how economic value gets extracted and how economic power works by creating software-based territories under the exclusive control of the platform owner, a control exercised through the management of data.[13] Not every attempt at platform expansion will work: Meta's current attempts to promote participation in the Metaverse right now seem to be leaving investors, shareholders and even its own employees unimpressed,[14] and no one knows at time of writing if Apple or Microsoft will fare any better. But that doesn't affect the general trend towards platform economies, which are expanding the territories from which data can be extracted in today's economies and societies.

There is also a third technique, which tries to extend the first (computer access) in order to gain the advantages of the second (platforms). It adapts physical spaces which we have no choice but to occupy (like our homes or our workplaces) and converts them into data territories via apps that, as we use them, automatically gather data from smart devices built into those spaces.

The Smart Home is basically a home reconfigured as a data territory. It is full of Smart Appliances, which each operate as secondary data territories. Their smartness benefits their producers more than us, since it enables them to sense how we interact with them and capture the resulting data for the corporation that installed them. We are talking here about the Internet of Things (or IoT), which has been hailed by business theorists as ushering in a new age of continuous connection. By contrast with the world in which advertisers tracked us only up to the moment they targeted us with an ad,

Smart objects can monitor us continuously as we use them, building an ever-richer picture of our habits for marketers.[15] In the Smart Home, avoiding being tracked is not an option: tracking is a basic feature of the data territory your home has become. Indeed, rent a property or buy a new home today and you may well be required to accept built-in Smart features, locking data territories into the larger order of home design.[16]

In the Smart Home nothing happens (no light is turned on, no front door is opened) unless mediated by a data-extracting app. Amazon's 2022 purchase of Smart vacuum cleaner maker iRobot for US $1.7 billion is a sign of where things are heading. After all, the Roomba (or at least the more expensive Roomba models sold by iRobot) literally can't work without turning the floorplan of your home into a data territory, from which data can be systematically extracted and monitored by an external corporation. The data that gets captured need not be narrowly defined. Amazon's separate home robot, Astro, has the ability to gather data just like an automated security guard, checking exits and so on, and then link up with your smart door bell, Amazon's Ring.[17] This is probably why iRobot's CEO announced after the acquisition that 'the home of the future is a robot'.[18] In October 2018 Amazon filed a patent for Alexa that would allow it to track emotions based on personal biometric data, signalling greater ambitions.[19] We'll come back to the Smart Home in Chapter Three, but who knows where the limits of Amazon's vision for its new data territories lie?

As in the early days of historical colonialism's territorial expansion, no one can be sure yet how much value will be extracted from such data: our argument does not depend, for example, on there being a thriving market for our Roomba data, or even on such data being shown on Amazon's balance sheet. It's the move to create data territories from which value can potentially be extracted that matters. That is the colonial move. The result is to transform something personal and intimate (the space of everyday life) into a terrain for exploitation by external corporations, making everything we do potentially legible to business.[20] Even our vacuum cleaners can leak images of us in intimate spaces to the world, as a reporter for *MIT Tech Review* discovered.[21] And even our own fridge can now be hacked. Indeed a whole new area called 'IoT forensics' has emerged, involving experts in extracting data from our devices.[22]

What will happen if we go on reorganising our homes and other spaces to become more reliant on devices, services and systems that transform them into data territories? The result, over time, will be societies that are also data colonies: networks of myriad data territories, each governed by the special power that data territories bestow on their owners. Business thinkers already celebrate this trend, without unpacking its social implications. 'Sensor data is here to stay', writes leading Big Data expert Thomas Davenport. The Internet of Things, together with Cloud Computing, Big Data and AI creates, according to Thomas Siebel, a 'Digital Transformation' that will revolutionise human knowledge and business.[23]

Common to all this is a vision so bold that, at first, it seems barely credible, except, that is, through a colonial lens. It's the idea that human life itself – the space of our lives – is now 'just there' for commercial use and exploitation in the form of data, a territory where business can 'write contracts', as Varian put it. Of course, this can't happen without major efforts of technological construction. The new colonial territories have to be explored (by writing the right code); and once identified, they must be expanded and exploited. In short, you have to build the 'data pipeline' as Varian once wrote, adopting without irony the language of one of historic colonialism's core assets.[24] Where is this pipeline plugged? Right into the flow of our daily life.

Data colonialism's process of relentless territorial expansion into daily life often comes disguised in the form of apparent benefits or even gifts (recall the 'gift' of electricity discussed in the Introduction). Not for nothing did Amazon decide early on to give away for free its smart assistant, Alexa. But it's important to see through that offer and focus on what, in a long-term historical perspective, is achieved by us accepting these gifts. We need to look, in other words, at the overall 'deal' of today's new colonialism. Its terms become clearer once we explore the implications of these new data territories for social power more generally.

New Data Relations Mean New Power Relations

Every data territory, as part of its role of maximising data extraction, draws us into a new type of social relation that we call *data relations*: relations between people and corporations or governments focussed entirely on maximising data extraction. Data relations work whether we know anything about data or not.

The first feature of data relations, considered from a social perspective, is asymmetry. Colonial relations historically were always asymmetrical: when colonial territories were acquired, the result was not free-flowing trade between colonisers and colonised, but one-way extraction of gold, silver, sugar, tea, and so on. When today you join a digital platform, the flow of data is equally asymmetrical: while the platform can see everything you do, you the user can see very little of what the platform does. Indeed, your data is analysed endlessly to 'personalise' your relationship to the platform. Perhaps you don't care about this, because that's exactly what you want (you *do* want TikTok videos that are what someone like you would want to see). But considered as a power relation, this is undeniably asymmetrical. It has taken years of social protest and legal battles to learn even a little about what platforms like Facebook and Instagram themselves do behind the scenes. Without brave whistle-blowers like Instagram data engineer Frances Haugen and Facebook data scientist Sophie Zhang, we would know even less.

In effect, data territories operate through a form of asymmetrical surveillance power. Such power seems less strange when we remember how such relations were a feature of colonial social life from common public spaces to slave plantations. Colonies were rarely democracies, and colonial powers ceded subjects only enough power to maintain legitimacy for their resource extraction.[25]

Data relations are not only asymmetrical; they involve (and this is their second colonial feature) new forms of power and control. Historical colonialism also extended the range of techniques whereby some humans control others: skin branding, the fingerprint and the security pass were all related to processes of colonial government. But because data – unlike land, minerals and crops – is itself a form of information, data colonies have some special possibilities.

Take Google once again. As Shoshana Zuboff (author of *The Age of Surveillance Capitalism*) explained, the vast amounts of data Google amassed by tracking how we search enabled it to predict what users will do next with considerable accuracy. Such predictions have value to other businesses who also want to know what we will do next (such as advertisers). Money can be made from this, for example, by selling advertising based on the assumed value of those predictions. But the absolute control of a platform territory allows its designers to go even further and, as Varian noted, experiment on us.

What might this mean in practice? Well, platform owners can nudge people in a certain direction and test out if they indeed follow a particular path; if not, then nudge them differently and track the new outcome, until you figure out the most effective nudging methods for a certain category of users. Multiply this millions of times across a platform and you get what legal scholar Karen Yeung called the 'hypernudge': a space of such pervasive influence that it becomes hard to know whether our actions there are free at all anymore.[26]

Far from slowing down, this process is accelerating. TikTok would never put it this way, but the skill of its algorithm is to convince us that what it selects for us to view is exactly what we would have wanted to see, even without TikTok.[27] Meanwhile, Facebook, under pressure since Apple in 2021 gave iPhone users the right to say no to being tracked by third-party advertisers, is now using its own AI to offer an experimental ad-testing service to those same advertisers, called Advantage+.[28]

We are not saying that platform owners literally want to govern every aspect of our lives, as an authoritarian government might. But the territory of the digital platform allows business to manage our actions in very significant ways for their own ends. Businesses call this process optimisation: optimising behaviour on the platform to ensure the best possible outcome for the platform's business goals. The drive towards optimisation characterises not just platforms and apps, but all areas of the internet where data is continuously captured for commercial ends. One thing that platforms always want to optimise is the amount of data they extract. It is not enough for a platform or app to occasionally monitor what we do; it must monitor us all the time. It is not enough for it to monitor just a few dimensions of our actions: it must monitor all relevant dimensions.[29] The possibility of 'N = all' in the jargon flows directly from the colonial nature of the data territory: nothing can happen there except what has been tracked as data.

Continuing with our review of the colonial features of data territories, their third key feature is an absence of the limitations that would normally apply to how power is exercised in a physical territory.

A data territory has no limits in terms of size. Whereas physical land and its resources are intrinsically limited by the surface of the planet, the territories under any platform owner's control can go on being created without limit, provided people can be persuaded to interact with them. In fact, expansion need not involve an entrepreneur creating new apps, but merely building links between existing platforms and apps. There is no limit to how one platform can link to other platforms or apps, or alternatively insert its own links into open online territory. Almost the whole internet is now completely 'platformised':[30] that is, effectively governed as data territories. You may notice this when big platforms act like gatekeepers to websites or smaller platforms, and you are asked to sign up via your Facebook or Google ID, or in China via your WeChat ID, ensuring that the data from you using that new platform is combined with your Facebook, Google or WeChat data, to those corporations' benefit. Underlying this may be more complex arrangements: Big Tech companies offering various data-related services to the smaller websites and platforms, which encourage them to build the large platforms into their own design.[31]

While 'platformisation' may sound like an innocuous descriptive term, looked at from the perspective of colonial history, it signals a transformation of open online space into further data territories for control and extraction.

Indeed, some platforms' territory is now so large that they are almost impossible to escape. If you live in China, you can do almost everything – buying food, getting loans, accessing government websites, checking if you can leave the house under the latest health restrictions – while staying on the WeChat platform. All the while, your dependence on WeChat and on WeChat's absolute power over your WeChat ID grows. There's an impact on business too: other businesses simply must maintain their presence on WeChat, if they want to remain in touch with you. Indeed, the Chinese government itself has special 'programs' on WeChat, interfaces that enable users to deal directly with government via the app. Perhaps this is the sort of one-stop app that Elon Musk has in mind when he talks elliptically of the super-app Project X. Whatever Musk's plans, China's super-apps offer us a glimpse of an all-encompassing economic, social and administrative power that even colonial power rarely attained, with the possible exception of Britain's East India Company.

Another way in which data territories are unconstrained is depth. There is no limit to the 'depth' at which data can be extracted:

as long as people are willing to go on interacting with the platform, the only limit is the code that platform designers write and the number of categories of data collection it installs.

This has huge implications for daily life. One is work, where, when tasks are translated into platform formats, managers are able to watch and track literally every level and aspect of the duties they assign to workers, including the time to the second it takes to perform those tasks. Another is the human body. We may have become used to the tracking of everything we do online, but we are much less used to the tracking of the human body. Facial recognition is controversial almost everywhere, but this is just the start of body tracking. A new industry is emerging around the tracking and analysis of how people talk, whether online or via their phones. CallMiner is a specialist company in the area of voice processing and its head has spoken of his vision 'to empower organizations to extract meaningful and actionable intelligence from their customers' conversations'.[32]

Data territories and their data outputs are also not limited in terms of their transferability. Whatever data is gathered via the data territory, its benefits can be transferred, shared or sold to others without restriction. Just as I can use information about today's weather at the very same time that you or the next person are using exactly the same information, so too can marketers with the dataset that defines you as a shopper. This is not to say that businesses don't sometimes create artificial scarcity around 'their' data. But from the perspective of platform users, it means that unless rules are written to stop this, they basically write a blank cheque when they give up data to any one platform. They have no control over where in future that data might move.

These features of data territories, while they might seem merely technical, amount to a redesign of social relations, or at least those social relations we perform inside data territories (already an awful lot). In those territories, we enter – seemingly voluntarily – into a series of asymmetric power relations, deeper and more pervasive than we knew before platforms were invented, through which businesses can optimise their goals without limitation. At the core of these new social relations is a new colonial reality: the capture of social territory on a vast scale for the control and benefit of the few.

Yet, as we'll see further in Chapter Three, this remarkable transformation of social life is presented to us – the users – as just businesses providing what we always wanted: more convenience, more social connection, and all in the interests of science, or at least 'knowing' us better. However, the engineers know such processes are anything but natural, even if they rarely own up to this. Which is why, when Andrew Bosworth, a leading Facebook engineer and now Meta's chief technology officer, admitted that 'the natural state of the world is *not* connected', he got a lot of attention in the tech industry.[33] People of course have always socialised and organised things together. But doing so by connecting to Facebook, and in ways that suit Facebook's business model, far from being natural, fits perfectly into a larger process of territorial capture.

Data, AI and the Environment

For data to be valuable, it needs to be processed, which consumes energy.

Our platform use and networking functioning relies on data processing by huge banks of computers (so-called 'data centres') which use electricity and generate heat, and so require cooling, as well as large amounts of land. Those processing costs get even higher for the most intensive data processing, such as large-scale AI. Aware of this, a number of Big Tech companies have green ambitions, with Apple aiming to be carbon neutral and Microsoft carbon negative by 2030. But the environmental costs of our digital life are still generally downplayed in public discussions.[34]

Consider data centres. They are distributed across both the Global North and the Global South. Most of us don't think about them, but their impact on the physical environment is starting to make the news. In west London, a mixed residential area where until a few years ago one of us lived, the building of much-needed new homes will be constrained at least until 2035 by the lack of electricity supply. Why? Because the area's vast expansion of data centres means there will not be 'sufficient capacity for a new [electricity] connection'. Globally, it is estimated that data centres will use between 3 per cent and 13 per cent of *all* electricity by the year 2030, compared to the 1 per cent they used in 2010.[35]

It gets worse. High electricity use in data servers causes overheating, which needs to be cooled down by using water – fresh water, it turns out. Just a month after that first announcement, Londoners learned that Thames Water, the city's water supplier, was concerned that its water supplies were getting dangerously depleted. One reason: data centres. A UK parliamentary investigation on the long-term implications of data centres on scarce water resources has been demanded.[36] And London is only one of 15 data centre hotspots across the world, not even the fastest-growing (Mumbai, Shanghai and Northern Virginia are outpacing it).[37]

The vast power usage required to 'mine' digital currencies is another area of headlong development with far too little consideration of environmental costs. 'Mining' here does not mean the physical excavation of resources in the ground, as in historic colonialism, but instead the use of vast computing power to validate the authenticity of each newly minted currency unit: for example, one Bitcoin. Bitcoin is just one cryptocurrency, currently in financial decline. But its 'mining' in 2020 used up more electricity than that consumed by Austria or Portugal, two medium-sized wealthy nations.[38] From the point of view of environmental costs, then, Bitcoin might as well be historical colonial loot.

Yet all of these consequences of computing and data for the natural environment are obscured by the abstract or metaphorical language used to describe them. For example, the 'cloud' on which our data is stored is not a reserve of water that can help sustain earthly resources, but a vast network of data centres that likely depletes them.[39] And this is only the data processing.

There are also the environmental costs of the physical *devices* on which we interact with data territories and data territories interact with us. All computing devices depend on chips which are generally made of silicon, phosphorus and many other minerals. Raw materials for these devices generally come from Africa, Asia and Latin America, with 36 per cent of the Earth's tin and 15 per cent of its silver going into electronics manufacturing.[40] Our devices also depend for their functioning on rare metals (such as cobalt and lithium) that are often mined in difficult and violent conditions in Africa and Latin America.[41]

Some of those raw materials require vast amounts of other natural resources to mine them. For example, lithium (used also in the production of batteries) is extracted, under one method, by pumping vast amounts of water to the surface in salt lakes, which then evaporates. In Chile's Atacama Desert this has caused water shortages for the Lickan Antay First Nation people who have lived for centuries on the land where mining goes on.[42]

At the other end of the process, there is the environmental waste into which our electronic devices turn when, all too frequently, they become obsolescent: each person on the planet is estimated to have generated 7.3 kilograms of electronic waste in 2019, according to the Global E-Waste Monitor.[43] Even though most of this waste is generated by the Global North, most of it ends up being disposed of in the Global South, with detrimental impacts to human health, including those who have to process it.

There is much more to say about the environmental costs of data territories, the colonial struggle for control over the resources that make our digital communications possible, and about the hidden environmental strategies of Big Tech companies to secure their power (for example by multiplying satellites that crowd Earth's thermosphere, or building cables that criss-cross our oceans).[44] But we hope the bigger picture is becoming clear. If the expansion of data territories represents a takeover of social life, the physical infrastructure that makes this takeover possible *itself* imposes major costs on the natural environment, especially in the Global South. This repeats a colonial history of the devastation of 'cheap nature', concentrated particularly in what Naomi Klein calls 'zones of extraction'.[45]

There's a Data Grab Happening (Very) Near You

Data territories can be installed in any area of life and exploit any number of informational dimensions.

One data territory which you may be familiar with is the one into which advertisers try to entice you when they track wherever you go, pulling together all the resulting data so that at any moment they can direct a relevant ad or other information to pop up on your screen. Such personalised marketing was an early sign of the tracking possibilities that Philip Agre noted three decades ago, but now it is facing some challenges. Not everyone in marketing thinks it has proven effective. Do *you* click on every ad that pops up on your screen, or even 5 per cent of them? In addition, some powerful business forces, like Apple (as already mentioned) and Google, have started outlawing the cookie-based tracking on which advertisers have relied. Sounds altruistic and privacy-conscious, although maybe Apple's and Google's goal is to boost the value of their own advertising and advertising-related data. It seems hardly a coincidence that a year after it forced advertisers to get explicit consent from iPhone users for tracking them, Apple announced a doubling of its own digital advertising workforce.[46] Some people, encouraged by this blocking of the tracking without consent that underlay Facebook's long-time business model, have seen in these developments the end of surveillance capitalism.[47] But data colonialism operates in many more spaces than targeted advertising.

Let's take personal finance, for example. Quite apart from the data scares involving online credit agencies, there has been a massive growth of algorithmically operated personal finance platforms and apps. These apps are certainly more convenient than waiting endlessly for a face-to-face appointment with a bank manager; you can do it all from your phone. But they depend not just on you supplying some initial data, but on you agreeing to the app accessing all sorts of data streams from you and third parties continuously, before and after the loan, to generate your constantly updated credit rating.[48]

Advertising and personal finance are just the tip of the data iceberg. Three other areas, basic to our quality of life, where you might not expect data extraction to be rife, but where in fact it is well under way, are agriculture, education and health.

All the Data That's Fit to Eat

Today, in many parts of the world, the tractor you see in a field is likely not only to have a computer built into it, but to be connected to systems for online data analysis in the cloud.

There is a large-scale drive towards 'precision agriculture', where US equipment maker John Deere and US AgTech corporation Monsanto (now part of the German Bayer Group) are big players. Although farmers are still free to drive and walk around their fields, their work may be largely prescribed by the software management programs that come with their equipment. For this purpose, genetically modified seeds count as equipment. To quote the FAQ of Monsanto's Technology Stewardship Agreement, 'any grower who wishes to purchase and plant seed that contains Monsanto Technology must have a valid TSA [Term of Service Agreement]'.[49]

This vision of precision agriculture is transforming animal farming too. China has been trying to expand its pig production for some years to meet huge global demand. Industrial-scale farming was happening without AI, of course, but the Chinese Big Tech giant Alibaba's Cloud business saw an opportunity. It offered its proprietary ET Agricultural Brain platform to monitor the progress of millions of pigs in huge industrial farms. First, however, it was necessary to train the algorithm to categorise pigs' behaviour and state of health correctly; for that, there are so-called 'digital towns' where migrant rural workers sit at computers all day performing such labour.[50] This is just one of many areas of food production in which Alibaba is involved.[51]

A new vision of food production is emerging that converges around data analytics, which Monsanto in 2013 called 'agriculture's next major growth frontier'.[52] Farmers contribute considerable labour, uploading and verifying images of their soil and weeds in order to train the algorithm. In many cases this is a one-way flow of information; the data collected by a John Deere tractor, for instance, is not accessible by the farmer who drives it.[53] Indeed, the farmer is locked in to John Deere's repair services, with no independent right to repair: a form of territorial capture that has recently been challenged by legislation in the US state of Colorado, which parallels challenges to Apple's decades of blocking owners' independent right to repair their devices.[54] In the process, a huge amount of data and power accumulates with the large corporations, while farmers become ever more dependent on management software such as Bayer/Monsanto's Field View platform (created originally by The Climate Corporation). As you'd expect, such services are targeted mainly at the largest farms, but they are resetting the whole food production industry. And there's another big player that benefits: the agricultural insurance business which can now use satellite imagery along with the new data platforms to monitor risks across large farms on an ongoing 'personalised' basis.[55] Just as Hal Varian predicted for the humble car insurance contract.

Farmers' relationship to their land, which they have managed for generations, is being transformed as a result of AI-driven platforms. This extends to the smallest scale: grass. The Grassometer, a device produced by former Irish company Monford, measures grass growth but a lot else besides. Via GPS and sensors, it maps the ground and measures grass height to bring 'precision farming to the grassland farmer'. Its founder proclaimed that 'data is the new soil: it's as important for the farm of the future as the tractor is today'.[56]

Although these developments have still much further to go, we can already see the territorial push to transform the old colonial territory of land into a new asset where data control is of primary importance. The physical territory of the farm and field is becoming a hybrid data territory.

The Classroom as Data Territory

Some of the largest players in Educational Technology (EdTech) are familiar names. Google Apps for Education are relied upon in many classrooms around the world, while Microsoft is also a huge player. Pearson, the world's leading education provider (who formed a major partnership with IBM in 2016), has a vision of a 'digital ocean', a boundless space where educational data is continuously generated from the EdTech-enabled classroom, just as social data is continuously generated from people's use of social media platforms.[57] The global EdTech market was valued at over US $250 billion in 2021, with predictions it will at least double by 2027.[58]

A core part of that value is data, data that can work well with the new large-scale AI services. As Pearson chief executive Andy Bird put it when asked if ChatGPT was a threat to its business, 'we are the owners of some very pure, rich data sets: when you start to input them into generative AI models, you get better outputs.' [59] Far from undermining a large EdTech player like Pearson, ChatGPT and generative AI enables it to extract even more value from its private data hoards.

Little of this is currently being regulated by governments. In fact, the education data sector is growing with the encouragement of governments in many countries. Brazil and Nigeria are examples of growing national EdTech sectors that are emerging with government support.[60] The Covid-19 pandemic gave a huge push to this. In South Korea, for example, all teaching went online at the start of the pandemic, with Korea's Educational Broadcasting System (EBS) playing a key role, but EBS in turn had deals with third-party commercial data extractors like Oracle's BlueKai.[61]

Indeed, the whole climate of EdTech's expansion has been set by multilateral organisations (that is, transnational organisations funded by multiple governments) that we might have hoped to be more critical of business values. However, from the World Bank and UN down, multilateral organisations seem to have fallen captive to the cheerleaders of the world's unfolding data grab.

Audrey Azoulay, the director-general of UNESCO, claimed in 2019 that 'education will be profoundly transformed by AI'. A UNESCO report from the same year talked of 'preparing learners to thrive in an AI-saturated future'. This fits with a broader business rhetoric that sees AI as automatically positive. The EU's High-Level Expert Group on AI stated that 'AI is . . . a promising means to increase human flourishing, thereby enhancing individual and societal wellbeing and the common good, as well as bringing progress and innovation'.[62] But that rather depends on how AI is implemented, doesn't it? If AI were really under the control of those impacted by it, that would be one thing, but under the rules of the data colony, the impacts are entirely different.

Public-private data partnerships are becoming the default way forward for education, as Big Tech seeks to make deals with governments, especially those that don't have much money to spend in education. Microsoft's 4Afrika is one such public-private partnership. Google Cloud is also offering higher-education support in eight African countries, while Facebook has recently developed an education app called Sabee, again targeted at Africa.

The lack of barriers to commercial values in the education space was celebrated in the 2021 report by Microsoft's Education division called *Understanding the New Learning Landscape*:

> Retail stores using AI can determine within 24h[ours] the impact on purchasing behaviors in response to moving the shelf location of a product. Education systems *should* be able to see the impact on student well-being and learning of changes in teaching strategies, tools and programs just as quickly.[63]

This sounds like Pearson's idea of the digital ocean. But the retail comparison raises a key question: is what needs attention in education as easily measurable as the rise in purchases when snacks are moved from one aisle to another? And what if it isn't? What happens when education's participants – teachers, parents, children – are regularly exposed to the asymmetrical power relations of the data colony?

An important recent report by the UK's Digital Futures Commission and 5Rights Foundation raised the alarm. It examined the data policies of two prominent EdTech platforms, Google Education and ClassDojo, intensively used in UK schools (with total UK downloads in 2021 in excess of two million).

Translating the report's diplomatic language only a little, four problems emerged clearly. Opacity: it's very hard to know what these platforms are doing with the data they collected. Expansionism: the platforms constantly blur the boundaries between school and home data, between core and 'additional' services. Unaccountability: the platforms have huge power, but their accountability to schools, parents and children is as yet unclear. And uncertain compliance: because they are so opaque, we don't know whether those platforms are complying with existing legal requirements on data (there have already been legal challenges concerning this, for example in Holland).[64] Nowhere in a school that operates Google Classroom or ClassDojo does there seem to be much scope for individual teachers to have a say in all this. How could they, when the contracts with Google Classroom or ClassDojo are negotiated at school level or even higher up the decision-chain? Parents have even less of a say, unless they are willing to withdraw their child from the school. And in case you didn't notice, the voice of the child, whose data is being harvested and archived, is completely absent here.[65]

Around the same time as the Digital Futures Commission investigation, the international organisation Human Rights Watch was surveying 163 EdTech platforms across nine countries. They found that government-backed EdTech during the pandemic involved multiple risks of data being transferred without consent to third parties, including in major markets like Brazil.[66]

Meanwhile, educational research in the digital classroom is proceeding apace. One area is eye-tracking. The hope is to build, as researchers put it, 'scalable attention-aware systems'.[67] Such systems would pay particular attention to a specific type of eye-movement that is evidence of 'mind-wandering'; a practice researchers say is worrying because it 'involves internal thoughts'.

Perhaps not coincidentally, behavioural management to control children's eye movements is catching on as a teaching practice. Some UK teachers, with support from the UK government, are championing an approach called SLANT. The words for which this acronym stands say it all: Sit up, Listen, Ask & answer, No talking, Track speaker.[68]

No one would pretend that, before the growth of EdTech, the classroom was an ideal space. It had many problems, particularly in low-income areas. But what we are saying is that, once you convert a classroom into a data territory, many older constraints on power disappear. Data territories, whether for teaching or any other purpose, are spaces of absolute rule that operate on scales and at depths unknown to the face-to-face teacher. This may be attractive to the rulers of large societies that need to plug huge skills gaps. Empires are being built in the EdTech sector based on aggressive marketing and continuous data extraction, like the Indian teaching app Byju's, which has raised more than US $22 billion in less than a decade, including from Silicon Valley (the Zuckerberg/Chan Foundation) and China (Tencent).[69] But Byju's is just one example of a vast global wave of corporate data extraction from schools and other educational institutions.[70]

Privacy and educational advocates are deeply concerned about a world where 'children's lives are filled with surveillance' to suit corporations who think that 'the more data the better'.[71] We must think about the long-term consequences for education itself: will it be organised increasingly around skills that are easy for EdTech to measure and monetise?[72]

My Health Data and Me

As we saw in Michael's story, there is no more intimate data for each of us than our health data. Used improperly or stored incompetently, and it could make the difference between life and death, or at least cause great inconvenience. That's why it is news when health data gets moved around without consent, as when DeepMind, an Alphabet subsidiary, collaborated with the Royal Free London NHS Foundation Trust and got access to large patient datasets without those patients being told.[73] It is even worse when health data – let alone mental-health data – is shared with social media companies, even if unwittingly, as happened with mental health startup Cerebral in March 2023.[74]

It's good therefore that many countries have laws that give special protection to health data, for example the European Union's

General Data Protection Regulation (GDPR). But there is some health data that is so sensitive – for example, an individual's genome – that philosophers argue that special rules are needed for how it is stored and accessed.[75] This is the controversial territory into which US gene-tracing apps like 23andMe have stumbled with what many regard as inadequate privacy protections.[76] It is a risk also in the area of mental-health data, a growing sector which has raised more than US $8 billion since 2020.[77]

The controversy over individual genetic data is just one small part of the landscape of data exploitation related to health. Societies and communities *do* need health data to protect us against disease. As the Covid-19 pandemic reminded us, disease spreads without regard for borders, and data about where and how it is spreading is essential to help us find cures or design treatments. Indeed, given the manifest inequalities of healthcare, a persuasive case can be made for more intensive gathering of national and international health data in anonymised form.[78]

But everything depends on how health data is gathered, stored and put to use – and by whom. Given that health data is individually sensitive, few would disagree with the principle of protecting its anonymity, outside times when it is being directly used to treat a patient. But, as in many other areas, experts worry that health data can be de-anonymised when it is combined with other data sources, and that's before we even consider whether the personal devices that gather this data are secure against hacking.[79]

There is also the question of the 'what'. Not only does the worry about anonymity intensify when we move from health data to the much larger and completely unregulated area of health-related data, but it is hard to know what health-related data is. Probably anything a commercial actor finds relevant to someone's health. But relevant for what purposes? An insurer providing a health policy would certainly like to know what we eat, how much we exercise and what we drink. Perhaps data relating to our mental health too, and any conditions that might affect it (marital unhappiness, for example), since mental health issues generally end up affecting physical health in the long run. And visions of 'precision medicine' encompass even individual genetic data.[80] That really is a lot of data! According to one estimate, the health Internet of Things market will be worth at least US $300 billion by 2025.[81]

In the past decade, a huge industry has emerged to exploit both our health-related and health data, with only the second being legally regulated. Countless app developers and device suppliers are competing to extract health-related data via our smart devices (Fitbit and Apple Watch being the best-known examples). This expansion is not limited to the Global North; India is now the world's biggest market for smart watches.[82] Big Tech companies are moving into the health sector, and capitalising on their Big Data, AI and cloud-computing resources: Amazon Health, Google Health, Microsoft Cloud for Healthcare. Amazon's proposed purchase of US primary care provider One Medical for US $3.7 billion positions it well as a potentially powerful player in the 'value-based care' market. This market rewards companies for patients' wellness, not just treatments delivered. But how will that be measured except by continuous data extraction?

A corporate move in November 2019 brought these two different worlds together and attracted a lot of attention: Google's purchase of Fitbit for US $2.1 billion. Why so much money? The assumption in the financial and tech press was that it was based on the value to Google of Fitbit's data and, even more important, Fitbit's potential as an expanding portal for health-data collection. As the *Wall Street Journal* reported, 'Fitbit . . . cuts out the middleman': a doctor or hospital are no longer needed for data to be contributed directly via Fitbit to a health-data corporation. When reporting the buyout, the newspaper could not resist a colonial metaphor: 'health services remain an open frontier.'[83]

As with education, the issues are larger than what happens to this or that piece of data. They concern the shifting balance of power in a health sector increasingly organised around the power that corporate access to ever larger health-data territories gives. This is the prize at which Big Tech players like Amazon, Apple and Google are aiming.[84]

The power play is however complicated by the role played by third-party users, such as insurers and employers. As Michael (in our opening story) realised, insurers are very interested in health-related data. And the employers who subsidise their workers' health insurance are themselves interested in the interest insurers take, because they need to keep down the cost of the policies they subsidise. So employers want their employees to wear the tracking devices that enable insurers to get a continuous stream of health data (in the way, for example, that UK insurer VitalityLife gets data directly from Apple Watch). These major commercial pressures are in tension with the important principles of protecting health data's anonymity and limiting the flow of data to minimise risks of de-anonymisation. The more companies interested in making profit from data enter the health industries, the greater the tension.

Meanwhile, the risks from losing protection of one's health data are very unequally distributed. Those who are poor and vulnerable will not be able to pay for legal action to secure protection of their data. On a global scale, countries in the Global South will have huge difficulties in securing protection against the free movement of health data when negotiating with the Big Tech corporations from the Global North. Big Tech's investments in health cover not just IT infrastructure, but direct healthcare, where they can offer themselves as solutions on the historically uneven distribution of health resources. Take the Chinese Big Tech platform Tencent, which is becoming a huge player in the health sector under the banner of 'precision medicine'. Tencent is expanding in AI-driven diagnosis and in direct healthcare provision via its Smart Hospital programme, with its social media platform WeChat being a key link.[85] Meanwhile, Chinese corporation BGI has emerged during the pandemic as a leader in Covid-19 testing and 'personalised medicine'; and in August 2022, ByteDance, TikTok's parent company, bought a major Chinese hospital company for US $1.5 billion.[86]

We need a new principle to strike a balance between individual needs for data protection and the social imperative to share some health data for the common good. A recent Commission on Governing Health Futures sponsored by the *Financial Times* and the *Lancet* framed this in terms of 'data solidarity'. They define it as 'simultaneously protecting individual rights, promoting the public good potential of such data, and building a culture of data justice and equity'.[87] We doubt whether this principle will be taken up by the international NGOs or national governments who set policy for the health-data sector. Fortunately, there are innovative social organisations like Hippo AI in Germany that are developing a different model for managing health data that builds on this principle of solidarity.[88] Without it, there is little to stop Big Tech players from further exploiting this health-data 'frontier'.

Data Territories and the Transformation of Work

The data territories of AgTech, EdTech and Health Tech may still seem distant from many people's lives, although they are a very pressing reality for those at the receiving end of these innovations. But there's one dimension of society where the innovations of data colonialism are already having a direct impact on people right across society. This is work and, in particular, the work of lower-status workers.

Work is being reconfigured by exactly the same mechanism we have argued is at the heart of data colonialism: the creation of new territories from which data can be continuously extracted and exploited. Low-skill work in particular is being transformed by digital platforms and interfaces of various sorts into forms of data territory under the absolute control of managers. As a result, whole ways of working are under threat.

There is nothing remotely surprising about capitalist or colonial employers grabbing all the information they can about their workers. As Karl Marx noted, surveillance is one of the basic tools of capitalist management,[89] while the unremitting surveillance and counting of bodies was a key feature of the pre-capitalist slave plantation and other sites of colonial exploitation.[90] So 'the quantified worker' (in legal scholar Ifeoma Ajunwa's phrase) is not new.[91] But the question is: how much worse are things getting through the new forces of data colonialism, which are making focussed and continuous data surveillance so much easier? The answer is: much worse.

In the most simple case, merely using your work computer can expose you to continuous surveillance by your employer, whether they are a thousand miles away or in the same building. So-called 'bossware' (platforms like Teramind, Veriato Vision and Clever-Control) locks workers into data territories from which there is no escape. It can track your every keystroke, look for certain keywords (which the employer may want to see being used), check in on you via screenshots of your desktop or video/audio recordings taken at intervals, or monitor the programs you are using to make sure you are not wasting time. This kind of system can also monitor work emails and your chat messages. The Covid-19 pandemic intensified the use of such techniques. According to one survey, 60 per cent of US employers were using bossware to monitor workers in 2021 (the figure was much higher in some industries like marketing and IT); another survey report expects 70 per cent of large US employers to be doing so by 2025.[92]

There are however other more subtle forms of computer-based surveillance that should not be forgotten. Take precarious low-paid workers, such as cleaners or domestic workers who are generally women. If their next job is just an SMS text or WhatsApp message away, having a phone connected to signal and Wi-Fi can well be the difference between getting work or not working at all (regularly missing that connection will be noticed by the employing platform and could result in getting kicked off).[93]

A great deal of low-paid work today is performed in data territories. The enhanced data-driven surveillance that results from this matters all the more, because of the profound reorganisation of work over the past four decades in many countries that goes far beyond the forces of data colonialism. Management experts now talk of the 'fissured workplace', as supply chains have lengthened and large corporations looked to reduce their direct employer responsibilities. As we would expect, the legacy of earlier colonialism shapes how this works in practice, with much of the lower-paid work being passed on to companies in the Global South.[94] More and more people now work further down the supply chain in precarious circumstances, where guaranteeing a steady supply of work means that a small company needs to comply with very specific standards for quality, timing and the like. Who imposes those standards? You guessed it: the large company at the top of the supply chain, such as Apple, Walmart or Tesco. And how are those standards monitored? Through the flow of data up and down the supply chain. We enter here the area of logistics, an area often ignored in discussions about Big Tech surveillance that focus mainly on social media.

The delivery warehouses in the logistics industry are a good example. We all want our Amazon package to arrive on time, but how often do we think about the individual human beings who are continuously tracked when we click on the icon 'track your package'? Here the territories for data extraction are generated not by interactive platforms that we play with in our leisure time, but by corporate IT systems. The systems that monitor how things and workers move around warehouses can be extraordinarily intrusive. The process of 'voice picking' (short for 'voice-directed order picking') involves using precise GPS readings of bodies and objects to monitor workers' movements and coordination from moment to moment. Its implications were noted more than a decade ago by the UK's GMB Union as a risk to workers' health, because of the unlimited intensity and depth of its surveillance, which pushes workers in a way that results in increased accidents and injuries.[95] On-the-body data collection leaves absolutely no space for negotiation, but even the seemingly less intrusive task of scanning itself for the workflow system is used without mercy: as Garfield Hylton, who works at Amazon's Coventry, UK, warehouse said, 'It's called "scanner adherence" – you have to be scanning every minute to show a constant rapid scan'. Otherwise, he said, a laptop-wielding manager comes to berate you.[96] But employees have no redress if the surveillance systems that monitor them get things wrong. Labour unions, among others, are increasingly concerned by this.[97]

On-the-body data collection is one thing, but the tracker need not be fixed to the body itself; it is enough for it to be fixed to the machine or vehicle that the worker is responsible for controlling. Truck-driving has been completely transformed by the insertion of devices into cabs which record how the lorry is driven, and report back to the employer. Electronic Logging Devices (ELDs) are now virtually obligatory in all US trucks. The change in power relations achieved by simply inserting such a device in a truck cab is profound, transforming the truck driver, as the US sociologist Karen Levy notes, from the sole story-teller of his time on the road to being just one information source for corporate databases among others.[98] The impact on drivers' sense of self-worth was profound. As one driver that responded to a regulatory consultation put it, 'I'm not going to work under conditions where I'm treated like a child, a child who doesn't have enough [sense] to know when to go to bed and when to get up.'[99]

Trucking firms, Levy reports, were also relying on data sources far from the truck cab; they were monitoring their drivers' social media use. Social media, after all, provide evidence to an employer of potential employability, or an employee's continuing fitness.[100] In the process, a key boundary between the worker's public and private life gets overridden. But the new rules of work-based data territory have potentially an even more dramatic impact: they can transform whole work sectors and even create new types of work that were not possible before, all through the flow of data.

The Gig Economy as Data Territory

We come here to the so-called 'gig economy': online task platforms such as Amazon Turk, RemoTasks and Upwork, driver platforms such as Uber and Didi (China), and food delivery platforms such as Deliveroo (UK) and Meituan (China). The leading Big Tech companies even have their own task platforms (Google's Raterhub and Microsoft's UHRS or Universal Human Relevance System). A more descriptive term for such tasks would be 'on-demand' work. The percentage of people getting work this way remains relatively low (under 10 per cent), at least in rich economies such as the USA and UK (although according to other estimates, freelancers will soon account for about half of the US workforce).[101] Outside the West, reliable figures are harder to come by. But the percentage is growing fast. In the on-demand work economy, platforms are the arena where work and tasks get distributed and monitored. As a result, they literally transform whole business sectors or create new ones. As leading Indian digital rights campaigners Anita Gurumurthy and Nandini Chami put it, platforms don't just dominate the market, they '*become* the market'.[102]

The owners of gig apps frequently claim their platforms benefit society. So, for example, transportation and ride-sharing platforms claim they improve driver wages and reduce the environmental impacts of driving, even though evidence may point to the opposite.[103] Meanwhile, the real advantages of gig platforms *for managers* derive from the dismantling of traditional employment patterns in the fissured workplace. Not only is the role of large-scale employers increasingly fragmented across long supply chains, but all sorts of employers (from the largest Big Tech platform to small companies lower down the supply chain) are increasingly outsourcing work they cannot do in-house. They can do this by recruiting on-demand workers across the world on a task-by-task basis who they reach via platforms such as Mturk, Cloudfactory, CrowdFlower and LeadGenius. These platforms give businesses access to large pools of workers, often in the Global South (for example, Colombia, India, the Philippines), who must perform precise tasks in fixed times and to fixed standards. This is where Uber drivers' photo-IDs get verified when their appearance changes and triggers a query from the system; and this is where small items of financial data get checked so that businesses can process it. On-demand platforms cover a wide spectrum of blue-collar and white-collar work.[104]

The basic features of such work follow from the properties of data territories explained earlier. Managers have always had considerable, sometimes even tyrannical, power over workspaces, but on-demand work platforms are designed to make sure that everything that goes on there is captured. The platform owner can change the rules from moment to moment as they please, including by denying a worker access to the platform, or arbitrarily paying them less.[105]

Ending someone's ability to work on a platform is now as simple as changing a few letters and numbers in a database. Literally, because workers for on-demand platforms aren't known by their names and faces, only by strings of letters and numbers.[106] And potentially legally, because platforms like Uber for a long time have been denying that they actually employ anyone, rather than just managing a database. Fortunately, law courts in various countries have begun to challenge this. Uber's denial of their employer status has now, in countries like the UK, been overturned; things are moving in that direction in the US too.[107] But this doesn't alter the pattern of power relations across data territories that the authors of a study on shopping apps call 'algorithmic despotism'.[108]

Such asymmetrical power relations now govern work for hundreds of millions of workers across the world. Systems of rule that are almost completely opaque afford very little, if any, dignity to their invisible workers. When gig platforms can lower the wages of workers who accept jobs quickly (because it is taken as evidence that they are more desperate), and raise the wages of workers who take longer to accept jobs (because they are not as desperate, and need enticement), we come to see what the system is designed to do: maximise profit at all cost, even by punishing the most vulnerable.[109] The result is to deepen the alienation of lower-status workers in society.

We are becoming ever more aware of the 'algorithmic cruelty' of work shaped by distant platforms that operate far from the human realities of workers' lives, and possibly with no human involvement at all.[110] So far it is mainly academic work and journalism from China and elsewhere that has reflected on what this is doing to the living conditions of workers. But British director Ken Loach's film *Sorry We Missed You* (2019) stands out as a rare exception to this near silence. It tells the story of Ricky, a driver, who is encouraged to give up a secure position with a delivery company and provide his services to a delivery platform which gives him no security and sets no limits to the disruptions that its data-driven tasks will impose on his family and mental health. The result is a nervous breakdown: an individual's whole way of life gets destroyed by the violence of automated exploitation.

Nor is dependence on gig economy work evenly distributed in either the Global North or the Global South: population groups who suffered from historical colonialism (including racial/ethnic minorities and immigrants) tend to make up higher percentages of 'ghost workers' (as gig workers are sometimes known) than other populations, at least in the US.[111] The detailed reality of the gig economy is therefore, to a significant degree, colonial. But so too is its overall organisation. Local ways of managing resources (often of course exploitative, but at least locally negotiable) are being replaced by large-scale global systems of management that have a new social reality at their core, the data territory under the absolute control of its creator.

Global Inequality, Redux

One thing is very clear: the inequalities of the old colonial order live on. We see it in the geographic location of the world's largest platforms and cloud service companies, or in the ownership of the undersea cables that transmit the signals we experience as the internet.[112] Or in the political relations that enable Big Tech companies based in the Global North to extract favours from the governments of the Global South. Or in the structure of global trade agreements that today impose the Global North's terms for data extraction on the rest of the world.[113] Yes, China is a newcomer to the colonial landscape, but, as arguably the world's oldest empire, it is behaving in no less a colonial way than the West in the era of Big Data.

As a result, the worst impacts of data colonialism are likely to be felt most painfully in the Global South and among those in the Global North whose forebears were on the losing side of historical colonialism. Tracing the intersections between the effects of historical colonialism and data colonialism will sometimes be hard.

It is as if two rivers, data colonialism and historic colonialism – one newly formed and the other very old – were running side by side. For now, we need to track both river flows and see how over time they will merge. But it is the new river's input to which we need to pay the most urgent attention, not because in the end it will matter more (who can say?), but because right now it is intensifying the flow of colonial history.

Chapter 3

DATA'S NEW CIVILISING MISSION

'I GUESS I'M frustrated because I thought the interview went well,' Angela said from her home in Manila.

'I'm sure it did,' said Phil from his office in Atlanta. 'But if what you are telling me is true, I wouldn't be surprised if the AI system wouldn't let the managers hire you.'

Angela was sure she was going to get the job, so when she received what seemed like an automated message informing her she didn't, she immediately called Phil to vent. Phil was married to her cousin, who lived in the US, and he had years of experience in HR and hiring people.

It's not like Angela was unqualified. Her previous job consisted of flagging hate speech in Filipino and English for a major social media platform. It was a stressful job, having to read all that filth. While it was not paid very well, at least it allowed her to work from home. Then her company implemented an AI system to detect hate speech in Filipino, making her skills no longer necessary. She had tried to convince her boss that the system was not very good, missing a lot of the language nuances. But the company needed to cut down costs. And so she was out of a job. She was now applying for similar positions in content moderation at different companies.

'I don't get it. The manager who interviewed me seemed to really like me. Can't he override the algorithm?' Angela asked Phil.

'I'm not even supposed to be discussing this with you, Angela. Our company doesn't like us talking about these details. But this company that interviewed you probably uses a similar system to the one we use, HirePlus AI.'

'So you are telling me the AI didn't like me?'

'It's like this,' said Phil. 'The HirePlus AI system scans your application materials, and also looks at your interview video, analysing your facial movements, word choice, etc. It's supposed to be very fair. It's better than being judged by a biased human being, trust me!'[1]

'But I have excellent qualifications, references . . . and I tried to be very positive during the interview,' Angela interjected.

'Yes, yes. I'm sure the interview was great. You have excellent presentation skills, Angie. I think the real problem might have been your survey responses.'

'You mean all those questionnaires I had to fill out before the interview? Those were very strange. They showed me a picture of some person's eyes, and I was supposed to determine whether the person looked angry, or passionate or whatever.'

'Those were personality tests,' replied Phil. 'They are meant to evaluate what kind of worker you will be. The hiring manager gets a score sheet that tells them where you fall in different metrics like introverted-extroverted, inflexible-adaptable, passive-aggressive, and so on. Only HirePlus knows how it all counts, but the AI makes the final recommendation, and that's that. Even if the interview team disagrees, they can't do anything about it.'

'Seems so unfair!'

'Well, maybe. But it's much more efficient. I attended a meeting where we were told the AI had recently been adjusted to ID workers who might show propensity to unionise, or to ask for raises.

Unions are a big worry for companies in the US right now, as you know,' Phil informed her.

'It just seems like it should be a human being making the final decision,' Angela said, trying not to sound too upset.

'I don't necessarily disagree, but that's just the way the world is right now, and we have to learn to adapt to it, Angie. Besides, think of how many more people the AI system allows us to interview that we couldn't interview before. That's a good thing, isn't it?'

'I suppose so,' said Angela dejectedly.

'Well, my advice,' said Phil, 'is that next time you take those interview surveys, try to answer in a manner that you think the AI will approve of, and not necessarily based on what you really feel. Anyway, I have to run now, Angie, but I want to wish you good luck. Keep us posted, OK? Someone will be fortunate to hire you!' And with that, Phil exited the call, leaving Angela to stare at a blank screen.

The Emperor's New 'Civilising' Clothes

Angela and Phil's story provides another example of how, for many people, the data grab is leading to a drastic worsening of the power relations that govern their lives. Yet, the narratives we are offered to explain or justify what's going on with data seem rather more comforting, hopeful even. Phil talks about the convenience of AI-based hiring, the ability to connect to so many more interviewees than could ever have been interviewed in person, and AI's superior judgement compared with humans. In this chapter, we'll unpack all those civilising narratives in more detail.

There is an overall similarity between the old and new civilising narratives that becomes clear only when we look through the lens of colonial history. In colonial history, the landgrab was justified by a whole series of narratives that were often very different from the realities on the ground. First, the evangelical goal of spreading Christianity, which for some was in direct tension with the brutal violence of the colonisers. Second, the idea of Europe's racial superiority and specifically its superior knowledge and science. Third, the idea of the colonisers' superior economic capacities and rationality (championed particularly by the British, who didn't take seriously the idea that the Pope had global authority to seize land in order to spread Christianity). Then there were elaborate mixtures of all three represented by narratives such as the West's 'manifest destiny', the 'white man's burden', and generally the idea of Europe's 'civilising mission' to transform the world through military conquest and socio-economic and cultural change. These narratives evolved successively over time to justify what was, at first, a bewilderingly new and large transformation of global power relations through land seizure. There were, to be fair, some early disputes among the colonisers about the justifications for their new-found wealth. But these were not serious challenges to colonialism, and they merely stimulated more sophisticated narratives of justification.[2]

Today's data grab is in its early years, and no doubt the narratives used to legitimise it will evolve too. But already there are clear patterns. No, proponents of Big Data and Artificial Intelligence are not saying that they want to civilise humanity. At least, not in those words exactly. The reason for this subterfuge is colonial history itself. During the struggles that led to the collapse of historical colonialism's political structures (for example, the British Empire in India and the French Empire in South-East Asia), civilising narratives as a whole became discredited. When Gandhi was asked what he thought about Western civilisation, he supposedly replied, 'I think it would be a good idea.'

So new narratives – or at least narratives that seem new – are needed to fulfil the same role that those disputed civilisational narratives once played. The advocates of large-scale data extraction use another language, invoking goals to which it is very hard to object, and claiming a very special practical power, as businesses or governments, to implement them. There's the idea heard in various circles of 'AI for Social Good'. There are narratives of supposed historical progress such as 'The Second Machine Age' or 'The Fourth Industrial Revolution' that are hard to challenge, if you make the assumption that history always moves forward in a linear way.[3] That idea, and the associated one that modernity is something 'the West' naturally leads, is itself a legacy of historic colonialism, as we saw in Chapter One.

But there are other civilising narratives on which the data grab relies which at first look entirely unrelated to the history of colonialism. There is the narrative of convenience and the narrative of connection, that is, the value of connecting people and things through technology. Both narratives, though superficially convincing, disguise the reality of the larger data grab and, in the case of connecting people, disguise some seriously toxic side-effects.

And the echoes from colonial history become even clearer when we unpack the detail: not only does the mission to connect promoted by someone like Mark Zuckerberg echo the religious mission of some historical colonisers, but, as we'll see in the chapter's last section, the language around Big Data and AI itself continues the vision of colonial science.

At their grandest, these stories express a civilising mission that makes the data grab seem the responsibility of a generous coloniser. Such grand narratives can be hard to unpack because they seem so enticing, so far beyond dispute. Yet they represent an ambitious rationale for data extraction, and they play a crucial part in how data colonialism works as a force in the world.

This process works on top of more obvious colonial continuities evident in Angela's story: economic dependence on distant US corporations has been a reality in the Philippines throughout its modern history, and now emerging tools like AI provide powerful new ways of managing life and work through data-driven structures of economic control that are truly global in scale, but that remain dominated by corporations in the Global North.

Not only is Angela being paid a small sum for insecure piece-work that involves being exposed to violent content, but when she loses her job doing that, she enters into a job-interviewing process where she is evaluated by an Artificial Intelligence program trained from data extracted from lots of people like her. Yes, there are humans still involved in the process, but more and more they play a secondary role, submitting to the recommendations of a black-box system whose authority Phil accepts without question, and Angela cannot contest. This huge transformation not only represents a continuation of very old neocolonial inequalities, but is evidence of distinctive *new* civilising narratives that work to legitimise the data grab wherever it is happening.

Data's civilising stories are another version of the Emperor's new clothes: empty in substance, but surprisingly effective in practice.

Civilising Narrative #1: Everyone Wants an Easier Life (aka Data Extraction as Convenience)

The first and most common version of Big Data's civilisational story is the argument of convenience. Convenience is always a societal bargain between needs that society requires to be fulfilled to maintain a certain way of life and other interests. What sort of bargain are we being encouraged to strike?

Smarter Lives?

We are often told that we give up our data in return for a service we like or need. But there's something strange about this supposed bargain. Although 'my' data is needed as one of billions of inputs to vast data platforms, by itself it's not valuable. Until, that is, it gets pooled into a huge vat where other people's data has been added, and the relations between all that data can start to emerge. This means that the real bargain we make is rather different. It is a societal bargain, not an individual one.[4] As societies, we are entitled to ask: is it a good bargain for all of us to enter into that arrangement? Is that really a good bargain at the societal level?

Certainly, modern life has become more complex, as we move between multiple online and onsite places for living, working and studying, and try to sustain contact with friends and family who are ever more scattered. Perhaps it is true that the tempo of modern life has sped up in the past couple of decades.[5] If so, maybe social media platforms' numerous shortcuts really do help us navigate these increased pressures. Think of the convenience of a WhatsApp group that allows users to share photos with their family or with multiple, but distinct, friendship circles, just by one or two clicks. Or the Facebook post that enables a hassled mother to publicise the details of her child's upcoming birthday party in one place for everyone to see, without numerous letters or phone calls. Or perhaps the digital personal assistant (or smart speaker) that enables you to open the front door or switch off the lights when otherwise engaged.

We don't deny the reality of these pressures and the usefulness of some of these solutions. But, given the huge landscape of data extraction we exposed in Chapter Two, we are entitled also to ask about the costs of those very same 'solutions' on a societal scale in comparison to the individual convenience they often provide. Three basic types of costs need to be considered.

First, there is the cost of the unequal power relations that result from the massive expansion of data territories in sectors such as agriculture, health and education, and over work and home life in general. As we saw, data relations are almost always asymmetrical, giving platforms and corporations a power to track you which they don't offer us back in return.

Second, there are the wider costs – indeed potentially serious harms – that flow from allowing large cross-societal systems of data collection to develop. We saw in Chapter Two the simple case of the car insurance contract which often now enables insurers to obtain not just the data drivers submit to them about their lives, alongside publicly available statistical data, but also a continuous data stream direct from their car that monitors them as they drive. The deal is sold as one of convenience, the convenience of lower premiums, though we suggested things might not prove so convenient, if, like everyone, you make an occasional mistake. But the long-term convenience is very much on the side of the insurance companies, since it fundamentally alters (in fact, narrows) the nature of the risks that insurers cover. What if health insurers start getting access to all sorts of other data they have not had before, for example, to our social media data or our search data? That will have even more dramatic implications.

Third, there is the risk that systematic data extraction enables completely new forms of extracting economic value and providing services. Here the risks may not be felt by the service user – Uber is undeniably convenient – but by those who work in the sector and now find their terms of work entirely dictated by distant platforms whose managers work in the Global North. These transformations are happening not just in the ride-share sector, but in other areas such as delivery, janitorial and low-skill jobs traditionally occupied by migrant workers. In these cases, convenience benefits some, maybe many, but at detriment to specific others. Once again, it is the arrangement (the enduring social and economic order) which can be built through data-rich platforms that should be our focus.

This takes us to a fourth and related cost, which is the effective and subtle reproduction of inequalities, which are the legacy of historical colonialism. As an example, consider home security. This is something everyone, rich or poor, wants and needs, for which well-off people anywhere will certainly pay, making this an area where Big Tech corporations present an attractive offer. But the introduction of these technologies into a society with deep histories of segregation, such as the US, is likely to be anything but neutral. Take Amazon's smart doorbell system, Ring, which is plugged into a wide network of law enforcement agencies. When Amazon bought Ring, it also purchased the Neighbors crime prevention network service (basically a platform for sharing reports of suspicious activity), and subsequently Ring has registered patents for facial recognition applications and other techniques. Facial recognition systems raise general questions of liberty, but their particular dangers for the populations who suffered worst under colonialism are also well known.[6]

Under the pretext of convenience, platforms want to become our social operating system. That way we'll use them more regularly and more intensively, and generate more ad income for the platform. Indeed, as the power of platforms in our lives grows, those platforms are able to position themselves as providing the *convenient* solutions even to the problems they help create. Do you want to protect yourself against 'revenge porn', the unauthorised distribution of your naked selfies on social media? Submit those very same selfies to Facebook! They will analyse their digital signature (presumably without a human being looking at them) and be able to detect and remove them when they are posted by anyone else.[7] Do you want to limit the kind of tracking that Google does? Buy a Google Pixel phone, 'secure to the core' and 'private by design'![8]

There's something else that is strange about our supposed bargain with Big Tech platforms: the timescale (is it really so convenient over the long-term?). Yes, I might right now, in the first flush of downloading a new platform, like its service. But later I may lose interest. That, however, doesn't make the data gathered from me obsolete or less valuable in the future. The data bargain, and particularly the societal bargain that comes from all of us making similar individual data bargains, are for the long-term – convenient for corporations, but not necessarily, long-term, for us. The resulting transformation of society's power relations, once under way, cannot be simply undone. So there's a real danger that the bargain we supposedly make trades individual short-term gain for long-term societal harm.[9]

Undeniably, the whole question of 'convenience' turns on other transformations of modern life that go far beyond the impact of digital platforms or data extraction: the decline of secure long-term work, the increasing need for flexible work hours to fit around complex patterns of childcare and care for elderly relatives, the decline of convenient physical locations that bring people together and promote information sharing (the pub, the community centre). There's an argument to be made, for sure, that social media platforms of many sorts – from Etsy and Instagram to Alipay and WeChat – play an important role in providing the resources we need to cope with increasingly insecure and complex lives. But are they the only platforms that could do this? Why would we not consider less data-extracting alternatives?

An Infrastructure for Living

The question of what is convenient takes on a special poignancy in Global South economies, which may have limited or damaged infrastructures. Convenience can quickly become necessity under those conditions. Even in one of Africa's richest countries, South Africa, the assumption that most people have a smartphone and a good phone plan does not apply.[10] For most of the population in Africa, in fact, using anything other than Facebook or WhatsApp for messaging or calls may be completely impossible, since Meta subsidises those costs as a way to get people to use its services.

In the Global South, then, Facebook is more of an enforced convenience, but one that, far from being natural, flows directly from the legacy of colonialism, and for two reasons. The first reason is the continued legacy of poor or absent telecommunications infrastructures linked to centuries of under-investment. The second, as we just indicated, is Facebook's privileged position of being able to negotiate with governments and telecommunications providers so-called 'zero-rated' deals for users to access Facebook (and a few other sites on which Facebook decides) at little or no cost. Facebook has negotiated these deals, known as Free Basics, with more than thirty African countries, plus many other countries in Asia, Latin America and the Pacific. This is not out of charity on Facebook's part, but a pragmatic response to its own stalling numbers of subscribers in the Global North. Indeed, since those deals, Facebook's total number of subscribers has started to fall, making its need for new subscribers elsewhere even more acute.[11]

When it comes to the kind of information-sharing that an economy demands, it is hard to argue against the perceived convenience of a service like Facebook in countries with poor transportation and poor communication channels for businesses. Some argue that, in Africa today, Facebook is the only place to get this kind of market information. Listen to Balqees Awad, a citizen of Sudan's capital Khartoum, speaking to Sudanese Guardian journalist Nesrine Malik about how she used Facebook in a period of acute shortages:

> When a bakery receives a bread delivery, or a petrol station replenishes its fuel, someone always posts in the group. They even tell us when there is heightened police presence in certain areas. Security patrols sometimes pick up people for no reason and extort or detain them.

A similar situation plays out for businesses trying to cope with an inadequate infrastructure to market their services. In Egypt, Malik argues, Facebook may be the only place to start a small business and test demand,[12] although it's important to note that in some parts of the Global South local networks are emerging that provide an information infrastructure for business without relying on Global North platforms like Facebook, for example India's Open Network for Digital Commerce.[13]

Is the strongest argument for Facebook's convenience that it supplies basic information infrastructure to countries already broken by the legacy of colonialism? If so, then thinking about Big Tech within a longer-term historical perspective requires us to think more about *for whom exactly* is the continuing social and economic inequality associated with Western platform power convenient? Maybe the interests of Facebook, a Big Tech business based in the Global North, are trumping the long-term interests of those societies' citizens. Within a colonial perspective, it is hardly new for colonisers to seek to govern through building convenient infrastructures.

Similar arguments can be made about the role that global media platforms like WhatsApp play in helping hundreds of millions of forced migrants to keep in touch as they cross borders in great danger and in search of better lives; or about the role of Facebook and Twitter in providing sites where political mobilisation and basic political discourse can happen at all in countries like Iran or Saudi Arabia that are deeply authoritarian. But we need to think a little more about the implications of such arguments. Are we really saying that Global North platforms work best in authoritarian countries with weak communications infrastructures, which are often former colonies of the West? If so, we are back to a colonial argument.

Meanwhile, it is also worth noting the increasingly important role that Chinese, not just US companies, are playing in Africa. It is not difficult to see the African consumer as a target for which both major powers of data colonialism are vying.

The narrative of convenience then, even if we are tempted to accept it, cannot be easily disentangled from both old and new colonial processes, and from the specific costs associated with the new landscape of one-sided data extraction that characterises many aspects of contemporary societies and economies. But let's turn now to the second candidate for a new 'civilising' narrative: connection.

Civilising Narrative #2: This Is How We Connect!

Everyone wants to be connected. At least, no one wants to live out their life in complete isolation.

Once again, digital platforms aim to persuade us that they offer the best way of realising a common need: this time, the human need for connection. Corporations like Meta or Tencent need us to believe that it's Instagram or WhatsApp or WeChat where we must go to be connected. To achieve that, they cloak their business models in all sorts of altruistic narratives.

Mark Zuckerberg, for instance, in February 2017, responded to worries that Facebook was becoming a site for dangerous populist rhetoric, such as that which fuelled the rise of Donald Trump. Let's put aside the new goals that Zuckerberg set out for Facebook in that statement – to support safer, more informed, more civically engaged, and inclusive communities – since they read like a shopping list for reversing exactly the problems that Facebook had been accused of accentuating. Let's focus instead on the overall vision offered by Zuckerberg in that statement, since it captures perfectly the civilisational narrative of 'connection'. And, as we read it, let's remember the religious dimension to many civilising narratives offered to justify historical colonialism.

> History is the story of how we've learned to come together in ever greater numbers – from tribes to cities to nations. At each step, we built social infrastructure like communities, media and governments to empower us to achieve things we couldn't on our own.
>
> Today, we are close to taking our next step. Our greatest opportunities are now global . . . Progress now requires humanity coming together not just as cities or nations, but also as a global community . . . Facebook stands for bringing us closer together and building a global community . . .

In times like these the most important thing we at Facebook can do is develop the social infrastructure to give people the power to build a global community that works for all of us.[14]

Two things are astonishing here.

First, the claim that a global community can be safely created through a platform with the exploitative business model of Meta. The hollowness of this claim was exposed a few years after Zuckerberg's statement by Frances Haugen, formerly of Instagram, in her testimony to the US Congress and European parliaments which revealed that Instagram had known from its own research the dangers its site posed, for example, for young women.[15] Haugen later commented: 'I did what I thought was necessary to save the lives of people, especially in the global south, who I think are being endangered by Facebook's pursuit of profits over people'.[16] Adam Mosseri, Instagram's head, at the height of the scandal responded: 'We know that more people die than would otherwise because of car accidents, but, by and large, cars create way more value in the world than they destroy. I think media is similar'.[17] As if Meta had no responsibility for its *specific* business model and its effects.

In any case, why believe that a global platform for humanity is even possible or desirable? But for Zuckerberg, it is impossible to think otherwise, because of his belief in the narrative of connection, a narrative over which Meta must remain in control.

The second thing that's extraordinary about Zuckerberg's claims is the idea that it is platforms like Facebook that uniquely represent the direction of history and of progress. That is a bizarre claim, when we consider that humanity has always been connected through small and large social networks. Did we really need Facebook to tell us that connection with friends and family is good?

Did humanity really need Mark Zuckerberg to make sense of its history?

The answer to both those questions is no – unless you believe that only Big Tech can bring 'connection' to humanity, precisely the sort of 'civilisational' mission that we are trying to unpack here.

Another version of this almost-messianic connection narrative predominates in the world not of everyday contact between friends and family, but between business and things. Data extraction from connected objects and devices is a key input to economic production, we are told – the form that technological momentum is taking in the twenty-first century – and none of us should stand in the way. As they say in the US, get with the program.

That's basically the argument of US West Coast tech evangelist and founder of *Wired* magazine, Kevin Kelly. His book *The Inevitable* was published, perhaps fortunately, just before the techlash kicked off in response to Facebook's 2018 Cambridge Analytica scandal. The book describes 12 'technological forces that will shape our future'. In reassuring language, it describes the predominance of surveillance and data extraction in contemporary capitalism, and why we should celebrate it. Three chapter titles are enough to give you the picture: 'Accessing', 'Sharing', and 'Tracking'. All, Kelly argues, are inevitable dimensions of humanity's future, a future Kelly welcomes rather than fears.[18]

But the idea that Big Data's data grab is inevitable has spread far beyond a few US West Coast evangelists. It is a tribal belief in much of Silicon Valley, well expressed for example by Silicon Valley entrepreneur Thomas Siebel's vision of a Digital Transformation that results from the convergence of 'cloud computing, big data, Internet of Things and Artificial Intelligence'.[19] And it is almost a religion in the wider global business community influenced by them. The pronouncements on Big Data from the Swiss-based World Economic Forum are full of such claims. Starting out from the cliché that data is the 'new oil' – 'a new type of raw material that's on a par with capital and labour' – the WEF conclude that hyperconnectivity and the data it generates will fundamentally transform human knowledge:

> All factors interact and co-evolve within an ICT ecosystem . . . a virtuous circle starts where improvements in one area affect and drive improvements in other areas . . . Individually, we are all limited in what we can know, but together hyperconnectivity makes it possible to overcome those individual limitations and mine different types of data to find insights.[20]

It's only reasonable to be suspicious about the frequent claims that only through continuous data extraction from people and things can life be connected in the way it should be.

Let's now look in more detail at these narratives, starting with the less familiar of the two, connections between things.

Connecting Every Thing

The Digital Transformation proposed by Siebel depends on connecting up people and things. For him, the Internet of Things (the world of 'Smart' tools and systems, all of them continuously connected to the internet) is much more than a marketing slogan; it is the building block of a new type of economy and society.

Starting with the intuitive idea of the Smart electricity grid – who, after all, would want a dumb grid? – Siebel swiftly moves on to the necessity, almost inevitability, of every object becoming connected, as well as human beings attaching objects to themselves in an attempt to become even more connected. As he enthuses, 'soon even humans will have tens or hundreds of ultra-low power computer wearables and implants continuously monitoring and regulating blood chemistry, blood pressure, pulse, temperature, and other metabolic signals':[21] Apple's iWatch and other fitness trackers already provide most of this.

Here, Siebel is expressing the general sense among Big Tech leaders that the Internet of Things represents the next new human–computer interface. In 2014, Cisco predicted that by 2022 connections between things would constitute 45 per cent of the world's internet connections.[22] While we don't know if that prediction came true, we do know that there are already more connected devices (13.1 billion) than people on the planet, and that number is predicted at least to double by the end of the decade.[23] To believe this is good for humanity (as opposed to just being good for corporations) depends once more on the civilisational narrative of connection, a narrative over which Western leaders increasingly want to retain control. According to the authoritative Council on Foreign Relations, and American think tank:

> As the Internet of Things (IoT) expands in coming years, the next iteration of the network will connect tens of billions of devices, *digitally binding* every aspect of day-to-day life, from heart monitors and refrigerators to traffic lights and agricultural methane emissions.[24]

This is a vision not of freedom so much as order, which is exactly how China's policy for the internet tends to characterise it. Indeed, China has been prioritising the Internet of Things for nearly a decade. Haier, a Chinese company, bought the US domestic appliance giant General Electric back in 2016 for US $6 billion, and China's latest five-year plan for 'National Informatisation' contains many references to building a 'sensory infrastructure' for the IoT, both in China and in other countries where China has influence.[25]

It remains to be seen how far such visions can be implemented, but according to this narrative, there is no option but for the whole world to become a data colony.[26]

Connecting Ourselves to Death

Looking back on the past two decades, it is hard not to recall the work of Karl Polanyi, a twentieth-century economic historian who wrote about how social relations had to be turned upside down to make room for the new types of economic relations that industrial capitalism required. Writing in the 1940s, after leaving Vienna for London a decade earlier due to the rise of the right in Austria, Polanyi's concern was the social conditions for the capitalist economic transformation of the nineteenth century. He described those conditions in terms of a gateway drug: 'highly artificial *stimulants* [were] administered to the body social'.[27] This artificial stimulation seems like an apt description for the impact that platforms like Facebook and TikTok are having on social life today.

Yes, the way that social media platforms connect us and stimulate our engagement is highly convenient for many things, including short-term political mobilisation. As a result, platforms like Facebook really have made a difference in countries with authoritarian political systems. In the early days of the 2011 Arab Spring, the availability of Facebook as a site where people could speak with apparent safety about common issues and find ways to mobilise did have an impact.[28] The same can be said of anonymous social media platforms that allow Turkish women, for example, to protest against the potential imposition of Islamic dress code by the state.[29] Kenyan writer Nanjala Nyabola strongly confirms this point, based on her extensive research into digital politics in Africa: 'there's no doubt in my mind that social networks have been useful for political discourse and for organising in countries where there is no free speech'.[30]

The value of digital networks *of some sort* – in the face of not just inadequate communications infrastructure, but political oppression – must then be taken very seriously. But, as with the narrative from convenience, we must weigh this connection narrative against other factors. First, it is just not true that social resistance never happened under pre-digital regimes: history is full of examples of people using what anthropologist James Scott once called 'weapons of the weak' (for example, agricultural or factory workers subtly working around their managers' strict rules).[31] Second, as many commentators on the Arab Spring pointed out, government authorities can not only use social media to promote their own ideas, but also to directly monitor protesting citizens. The failure of any enduring revolution to emerge in Arab countries after the 2011 Arab political revolution was probably not unconnected to this kind of surveillance, as well as to weaknesses in the kind of politics that are possible on social media.[32]

The social side-effects of social media platforms must be considered seriously. From the history of colonialism we know that side-effects matter. When historical colonisers intended to invade and take the gold, they didn't know (at first) that they were bringing European diseases which killed far more than their weapons ever did. Today's data colonisers certainly intend to seize data territories and get control of the resulting data. But no one is saying that they predicted, let alone planned, the side-effects of their platforms and business models on the social world, which many claim have toxified social life.

Many now argue that Big Tech's hunger for profits and their lack of concern for the very people who make their wealth possible have contributed to the creation of a deeply fractured public sphere, where 'connection' is divisive, exhausting and often traumatising. But we should keep things in perspective and remember that the origins of cultural, social and political polarisation long predate Big Tech: it makes no sense to say that digital platforms are the *prime* cause of polarisation.[33]

At the same time, the ability to join virtual groups who share our interests – the feature we were all excited about in the early days of the internet, before it was thoroughly commercialised by Big

Tech – is the very same feature that some claim has created 'filter bubbles' where disinformation and hate thrive. A term originally proposed by Eli Pariser in 2011,[34] a filter bubble is basically an internet-era updated version of an echo chamber: a space where you only hear what already fits your way of speaking and thinking. The idea is that search engines and social media platforms, by doing what they do best – that is, giving us the content that we want to see, and eliminating the content we don't – create unhealthy diets of information that reinforce our existing worldviews, rather than presenting information that might challenge them.

In other words, connection, as operationalised by social media algorithms, has become something extremely targeted, something that limits, rather than enhances, diversity. Platforms, not people, decide which connections are more advantageous to them: that is, whatever gets us to spend more time on the platform, as epitomised by TikTok's 'For You' algorithm. The more time we spend on platforms, after all, the more data that can be extracted from us.

While there's no denying that positive movements for justice (such as Black Twitter in the US)[35] have emerged from digital platforms, there remain a number of negative consequences to be considered, which flow directly from social media's business models that are based on data extraction. According to their critics, filter bubbles and echo chambers have resulted in everything from Brexit to the election of Trump, Bolsonaro and Duterte, the antivaxxer movement, global-warming deniers, and so on. Every bad idea in the world is given a platform to thrive thanks to filter bubbles, and while good ideas should enjoy the same virality, the theory goes, they are always somehow overshadowed by the angry voices. According to US commentator Jonathan Haidt, this is because people, now armed with the weapons of the Like, Share, and Retweet button, are more willing to promote content that triggers their outrage. The result has less to do with authentic connection and more to do with public shaming and mob mentality, 'amplifying' in the words of leading US legal scholar Julie Cohen our 'collective unreason'.[36] But these social media features (Like, Share, RT) are themselves just techniques for social media platforms to stimulate us to generate ever more data about our 'engagement'.

When we keep our focus on the wider impact of social media platforms' business models, it matters less that, contrary to popular belief, academics have found scant empirical evidence that filter bubbles actually exist, at least as a widespread social phenomenon.[37] People who spend time online are generally exposed to more diverse content than the people who don't. But is polarisation in information flows the only type of polarisation that matters? What about polarisation in how people feel about each other – those they seem to be like and those who seem different from them? Political scientists call this 'affective polarisation'.[38] Isn't this more consequential in the end, including for how people feel about the news content they come across?

The journalist and Nobel Peace Prize winner Maria Ressa has experienced first-hand the power that social media algorithms, designed to prioritise profits over facts, have had on polarisation and democracy's death by a thousand 'likes'. As president and co-founder of *Rappler*, an online news site in her native Philippines (a nation whose population spends the most time on social media in the world), she was probably the first to present to Facebook incontrovertible evidence of how their business model was allowing for the commercialisation of disinformation, only to have that evidence ignored. Reflecting on what she learned from the experience, she writes: 'Tech sucked up our personal experiences and data, organized it with artificial intelligence, manipulated us with it, and created behavior at a scale that brought out the worst in humanity.'[39]

The impacts can be devastating. Social media platforms really can have a direct effect on the extremist radicalisation that is part of social polarisation.[40] One example, documented in a *New York Times* podcast series,[41] involves the case of Caleb Cain, a US high school student of progressive and liberal sensibilities who, in a two-year period of watching thousands and thousands of YouTube videos, went down a Rabbit Hole (the name of the podcast) of increasingly extremist content, and ended up identifying as a white supremacist. Fortunately, he later made the reverse trip and ended up watching progressive content again and renouncing his far-right allegiances.

What's interesting about cases like this is that the reason for this push to extremism seems to be a simple change in YouTube's recommendation algorithm around 2015 (just to be clear, similar recommendation algorithms can be found in other platforms, so this is not a problem unique to YouTube).[42] Such changes illustrate the huge power that accrues to owners of data territories. Instead of showing users more of the same content they are interested in, social media platforms have figured out that the way to get them to spend more time on their platforms (and thus to be exposed to more advertising) was to show them new things they might be interested in. For cat-lovers, not more and more cat videos (how many in the end can they watch?) but, say, videos of tigers, since tigers are also felines; that way, users will spend even more time on the platform discovering new content.

If only we were talking about just cats and tigers! The kind of algorithmic homophily (birds of a feather sticking together) that platforms encourage gives rise to online spaces where far-right extremism can easily flourish because already existing polarisation gets focussed and amplified there.[43] We saw an example of this during the 6 January 2021 insurrection in the US, when a group of people living in such a radicalised bubble collectively decided it was a good idea to storm the Capitol building. But countless other examples can be collected from around the world, from Brazil, India, Kenya, Myanmar and Sweden, to name just a few.

Did far-right sentiments and opinions exist before the internet? Certainly. Do platforms like Facebook encourage people to become more radical, finding those you like and pushing each other into committing ever more radical acts? Yes, there is overwhelming evidence from political scientists that such polarisation and radicalisation are going on.[44] Even worse, there is evidence from whistle-blower Sophie Zhang that Meta *knows* this is a consequence of their business models.[45] And have companies like Meta and Google benefitted financially by allowing climate-deniers, antivaxxers, election-deniers, and all sorts of conspiracy theorists to advertise on their platform? Yes, they have.[46]

It is clear we all want to be connected – but not necessarily in this way! Not if 'connection' means limited templates for human interaction determined by business models. Not if 'connection' means more data extracted from us for profit, with more of our gadgets constantly surveilling us in order to use that information to sell us more junk, or to manipulate us. Not if 'connection' actually means its opposite: an easy way to end up disconnected from wider reality and from people we might usefully disagree with. In short, not if communication – the basis for connection – means not the sharing of meaning, but the gamification of meaning, where what counts as truth is established not through discussion, debate or even political negotiation, but merely through the virality of content.[47]

Legislative moves are under way in many countries (for example, the UK's Online Safety Bill, Brazil's Fake News Bill,[48] and multiple proposed bills in the US) that may impose stricter responsibilities on platforms. But none so far appear to challenge the basic business models of data-extracting platforms. The result is that, even if we sometimes feel digital platforms are places for fun, the real game is likely to remain the one that is being played *on us* – by the algorithms and the large systems they serve.

Which brings us to algorithms and the last important civilising narrative we'll discuss: the story of extractivist science and knowledge associated with AI.

Civilising Narrative #3: AI Is Smarter than Humans

We have all been exposed to the idea that Artificial Intelligence is, supposedly, a great thing for humanity. Narratives about AI and machines that can think like humans have been around in science fiction and other forms of storytelling for a long time. We are waiting – in hope or in fear – for the day when we can really talk to a machine like we talk to another human being, or the day when we can rely on a self-taught robot to perform unimaginably complex tasks flawlessly and efficiently, liberating us from the drudgery of work and from the follies of our own biases and limitations.

What we have today with Artificial Intelligence is often not very intelligent at all: it is what AI expert Meredith Broussard calls 'artificial *un*intelligence'. [49] First, there is the problem of training data: the repetitive calculations on which AI depends only work because it has been exposed to thousands if not millions of examples from which it can 'learn'. AI programs need to cut their teeth, as it were, on preselected datasets. But those datasets necessarily have limitations: they only include things that have been previously counted. Whatever is counted necessarily reflects the forces that shape what there is in the world, forces like privilege and inequality (which is why AI is often found to replicate the biases that are already found in the world).

But a second problem arises with the most advanced forms of AI, the so-called AGI (Artificial General Intelligence). These systems are able to move beyond their original task parameters and come up with innovative solutions, going as far as potentially generating new knowledge. But the problem of context remains. Human knowledge arises in a life context, many aspects of which are so obvious that they never need to be made explicit – at least to humans. But to AI, since it is formed outside of any life context, everything needs to be made explicit, including the unacceptability of what humans would reject as nonsensical. All the recent absurdities thrown up by our questions to ChatGPT are variations on this problem of context.

Beneath these examples of AI failures is a more fundamental point about what AI actually is: a mere collection of programs that are given massive amounts of data, find patterns in that data, and make some statistical predictions based on those patterns, such as predicting what the next word in a sentence is likely to be. That is true of small AI just as much as it is of the largest AI, for example the models for predicting text that drive the now famous ChatGPT or Google equivalent system, Bard, as even the most enthusiastic celebrants of AI have to admit.[50] And there lies the problem. Let's quote the academic paper that lost leading AI researcher Timnit Gebru her job at Google:

> An LM [AI-driven language model] is a system for haphazardly stitching together sequences of linguistic forms it has observed in its vast training data, according to probabilistic information about how they combine, but without any reference to meaning: a stochastic parrot.[51]

The problem occurs when the parrot embarrassingly repeats bad things it has heard. For example, researchers discovered that when asked to adopt certain identities ('You are a 20 year old White male with a high school diploma, earning US $30,000 per year. You are a registered Republican living in the USA.'), ChatGPT produced racist and polarising ratings of various groups that were seven times larger than the average ratings produced by actual humans.[52] We are not denying that there can be real value in using AI programs, like when they are used to design vaccines, detect cancer or model environmental phenomena; that is, where AI can see into datasets much too large for human intelligence to map and find statistical patterns. But major problems arise when we unquestioningly use the outputs generated by AI to simulate human understanding or make decisions that impact human welfare.

Here is where the civilising narrative comes in, as a way to influence our judgement about what is best, and to justify why data for AI needs to be extracted from us in the first place. This civilising narrative is complex, and it will take a bit of work to explain why it is continuous with the even larger story of colonial science and the interconnections between colonialism and modernity that we have been exploring.

Science and Power from Colonialism to Capitalism

AI is part of science. It is a discipline within computer science, and while most people don't think of computer science in the same way they think of 'pure' sciences like biology or physics, it is a science, nonetheless. We discussed in Chapter One the important continuities in the evolution of scientific methods, from early colonialism until today, and what those continuities have to do with power.

Francis Bacon, considered one of the founders of the scientific method, established in the seventeenth century a direct connection between the growth of science and the exploration of new worlds: 'Nor must it go for nothing that by the distant voyages and travels which have become frequent in our times many things in nature have been laid open and discovered which may let in new light upon philosophy.'[53] Since Bacon was writing in 1620, we know that the distant voyages and travels he mentions were the explorations happening within the context of full-scale colonial expansion. This suggests that the scientific method and practices of colonising the world were, from the start, joined at the hip.

More precisely, both scientific and technological evolution are part of a colonial impulse to administer extracted resources. The scientific method and colonialism developed and evolved together, in response to the problem of managing what Bacon called the 'many things in nature' that 'have been laid open'. The need to acquire and transport resources more efficiently, to rule new territories across huge distances where those resources were located, and to process those resources into something that generated wealth, were all colonial problems that required greater scientific and technological advancements.

Almost three hundred years later, during a lecture to the Liverpool Chamber of Commerce in December 1899, Sir Ronald Ross shared the results of his ongoing work on combatting malaria (which killed many English colonists) by stating that 'in the coming century, the success of imperialism will depend largely upon success with the microscope'.[54] This statement helps to illustrate the link – already present in Bacon – between the colonial enterprise and the scientific method, which was used as a way to reduce – or abstract – the natural world down to objects that could be managed and properties that could be predicted. From this perspective, the scientific method becomes not only a tool for understanding the world, but a process for eliminating everything that interferes with its colonial management.

This process had a significant impact in understanding the diversity of the colonised world not as a positive feature, but as a problem to be dealt with. In the words of postcolonial critic and Goan environmentalist Claude Alvares, '[a]bstraction increases control by homogenizing its subject matter. It eliminates the basis of diversity, the personal and the historic, creating an artificial reality which can be completely controlled'.[55] This is probably why many feminist critics of science have also noted the close ties between Bacon's metaphors for scientific method and the masculine imagination.[56]

Just to be clear: we are not suggesting that every and all scientific discoveries serve a dark colonial purpose, or that we would be better off without science. Our point is merely that there is an undeniable arc in modern science, starting with colonialism, that bends towards the goal of controlling people and nature, and that links scientific abstraction and indeed technological development to extractivism. Our concern is that this arc extends all the way to the science behind AI.

The Discriminatory Intelligence behind AI

Of course, most of the time we don't know how AIs work, because their inner works are black boxes protected by their makers as trade secrets. And it's not like their makers always understand how their programs work either, because AIs can generate outputs on their own, without human intervention. In the case of AIs like the one described in Angela's scenario, we imagine that the program can somehow absorb the collective experiences of, say, thousands of expert human-resources managers and interviewers, and use that knowledge to act like a super-human recruiter that can expertly and quickly assess job candidates, determining who is the best person for a job while avoiding all the biases and inefficiencies of humans. This is the narrative that Angela was asked to accept. But as Dan McQuillan observes, AIs are simply programs for making 'elaborate statistical guesswork', based on extracting data that, in turn, reflects the pre-existing inequalities in society.[57] Sure, that's a pretty neat technological trick in and of itself. But the problem, he argues, is that this form of statistical optimisation is being used blindly to reinforce forms of discrimination that often go back to colonialism, all while being presented as a magical new solution to social problems.

To understand this, let's break down Angela's example a bit more. The experience we described was based on what we know from how companies like HireVue and Humanly market their products, or from what we hear from people who have been interviewed using such systems.[58] But we can infer that, in order for the AI to determine whether Angela was a good candidate or not, it didn't need to learn human resource policies, it didn't need to learn effective interviewing techniques, and it didn't even need to learn about the qualifications for the job. Instead, the AI was given tons and tons of CVs, survey results, and videos of actual interviews (what is called 'training data'). Then it was told which interviews resulted in hires. The AI then compared the 'good' interviews and the 'bad' interviews and identified the characteristics that occurred more frequently in the good ones (the characteristics that were more statistically significant, in other words). The more training data the AI is fed, the better, because it can find some unexpected correlations that humans (whose capacity for processing large amounts of data is limited) could not find.

The science and maths behind all of this is fascinating. But it is already clear where this model soon risks going wrong. Maybe the AI determined that good candidates were the ones with more years of experience, or those who used certain words more often, or those who answered in a survey that being on time is more important than stopping to help someone on the way to work. But maybe it also learned that in addition to these characteristics, a good candidate is also . . . male and white. In other words, if the training data reflected the fact that more white males ended up getting hired, and the AI was not explicitly adjusted to correct for this, it could have 'learned' that being a white male was a characteristic of good hires. To further complicate things, all of these decisions may not be apparent to the people who designed the AI, or who are using it. They just receive the final recommendation or prediction, and it takes some extra work to figure out if the AI is biased. In one recent case, Amazon decided to abandon plans for a recruiting AI system precisely because, in testing, the system was failing to recommend hiring many women (Amazon never disclosed why it didn't work).[59]

The people who design and sell these systems will probably complain that our example is too simplistic. They will also point out that even when AI is shown to exhibit some biases, those can be identified and corrected, which is not something we can do easily with human bias; in other words, they would say, AI represents a better chance at eliminating bias. They would also probably point out that, while we wait for these systems to be improved, we can simply include a human being in the decision loop to ensure proper outcomes.

In response we would argue that the fact some of these systems are discriminatory and opaque means that they are probably working exactly as intended. AI is applied in all sorts of areas today, but the most concerning application – and the one we can expect to see increasing – is to automate decisions that have a negative impact on populations that can do little to contest these outcomes. These populations are the same ones who have traditionally suffered discrimination under violent and bureaucratic systems, and who are on the 'wrong' side of the intersections of class, gender and race. So yes, these systems are very convenient, if by convenience we mean, as McQuillan argues, 'not having to engage with structural legacies of colonialism'. Simply adding a human in the loop will only serve to locate an individual scapegoat to blame when something is found to go wrong.[60]

More important than these limited corrective mechanisms (which have only emerged in response to recent outcry) is the bigger picture: that most applications of AI are part of a colonial history of scientific and technological development that, as scholar Paola Ricaurte argues, uses extracted data to codify and automate already existing violence and discrimination.[61]

AI as a Colonial Tool of Extraction

We can probably assume that, working in their research labs, data and computer scientists are probably not thinking about creating what mathematician and data scientist Cathy O'Neil calls *weapons of math destruction*. [62] And yet their scientific methods, and the tools created with those methods, are part of a long trajectory of attempts to manage the world more effectively and more profitably. Data has been, since the beginning of colonialism, an important tool to extract the 'many things in nature that have been laid open', to cite Bacon again.

Against this background, we can see clearly that AI and particularly large-scale AGI is just the latest step in the colonial production of knowledge. The basic reason is simple, even if its implementation is very complex. As we noted right at the start of the book, large-scale AI takes the whole of humanity's production that is registered online as its input. From this, it seeks to generate knowledge that it can offer back to us as valuable. Doing this requires vast computing power, since we are talking trillions of data points, possibly orders of magnitude more. And that computing power remains the exclusive possession of the largest Big Tech players. This huge computing power is transforming the world of search engines, but also how social media platforms like Facebook are generating useful targeting for advertisers via its AI-based Advantage+ program developed for advertisers as noted in Chapter Two.[63] Whatever the details, the very idea of treating human culture as if it were just free input to AI processing is, from the start, one more act of colonial appropriation.

The second colonial feature of AI today flows from the basic point that AI needs training data to function. The task of collecting and organising this training data into a format AI can use, far from being 'artificial', is a labour-intensive process with lots of human input. As a collective of workers for Mechanical Turk puts it, 'There is no ChatGPT without human workers'.[64] This labour is very often done by poor people in the Global South, an unequal relation that replays the global division of labour that is the long-term legacy of colonialism.

Companies have realised that people in the Global South who are willing to work without contracts for reduced wages in precarious conditions form an ideal labour market to do these repetitive tasks. As we saw in Chapter Two, platforms have been created to control these labour markets. Leading tech journalist Karen Hao tells the story of a gig worker in crisis-ridden Venezuela who works for a company called Appen labelling images, making as little as US $6–$8 a week, which she can't even withdraw from her account because the company has a policy of only allowing withdrawals when workers accumulate more than US $10.[65] Those gig workers may be highly educated, but they are also, as recent research in Brazil discovered, likely to be women who lack secure employment and so be willing to accept extremely low wages.[66]

AI industries and their contractors have no qualms recruiting remote workers who live in conflict zones like Venezuela or Syria, which sounds positive until we realise the goal is not to provide good jobs to desperate people, but to take advantage of their vulnerability to create pools of workers willing to work for very little. These jobs often entail being exposed to violent content, as with the Meta contract workers in Kenya who are given fifty seconds to determine whether there is violent content in an uploaded video, regardless of the length of the video or how traumatising it might be to watch it.[67] Those same workers have sued Meta in a series of law cases that, at time of writing, are still unresolved. Without realising it, Angela was probably training the very same AI hate-speech detection system that made her redundant: that, after all, is the point of this human labour, to train AI to replace it. The fact that an industry selling us a vision of an autonomous intelligence for solving the world's problems relies on armies of underpaid and exploited workers in the Global South is one of its biggest dirty secrets.

The third colonial dimension of AI relates to how its techniques of data processing operate in detail. Take the question of the languages that provide the input to AI. Almost all of us benefit from AI being able to recognise our voices or translate from one language to another easily and accurately, through a process known as Natural Language Processing (NLP). But what happens when the uneven application of these resources reinforces colonial legacies? What happens, for instance, when Big Tech decides to prioritise the development of NLP technologies in the languages of the colonisers (English, French, etc.), not the colonised? The world watched with concern when Facebook's inability to properly monitor hate speech in Myanmar allegedly aided a genocide (Facebook's engagement-based algorithms promoted inflammatory content advocating discrimination and violence because such content was calculated to keep people in the platform for longer, generating more advertising revenue). Algorithms that monitor content in majority languages like English and French have more mechanisms to prevent this from happening, but the question of Facebook's responsibility towards its speakers of minority languages, like Burmese, remains.[68]

Alternatively, what happens when AI discriminates against the language of the colonised? A report prepared by Business for Social Responsibility (BSR.org) found that during a May 2021 conflict, Arabic content published by Palestinians across Meta's platforms was censored to a larger extent than content in Hebrew published by Israelis, even when the latter contained clear examples of hate speech and incitements to violence (it's possible that humans, not just algorithms, played a part in this censorship).[69] In an earlier case in 2017, a Palestinian man was detained by Israeli police after he posted an Arabic message on Facebook that read 'good morning', but which was erroneously translated as 'attack them' in Hebrew.[70]

Some of these problems have been in the public eye for a few years and Big Tech companies have started to respond to them (for example, the egregious faults of Google's AI that had coded Black bodies in deeply negative ways).[71] But such responses to AI's problems have been far from complete, which takes us to the fourth way in which AI operates as a tool of colonial power: the highly unrepresentative consultation processes that supposedly try to correct for AI's problems.

Those doing AI research are overwhelmingly from the Global North, and more likely to be men than women. Take the specific case of NLP research: as the 2021 State of AI Report makes clear, NLP research is becoming the domain of a handful of Big Tech corporations and their university partners in the Global North, perpetuating the very same inequalities that contributed to such problems being undetectable or unaddressed.[72]

Unsurprisingly, industry-driven solutions to issues of algorithmic injustice and AI unfairness tend to be weak. AI Ethics is a big field these days, largely funded by Big Tech. According to digital cultures researcher Luke Munn, it is also useless, for three reasons.[73] The principles it proposes are too abstract to be of value. For example, the University of Montreal issued a declaration which says, in part, that AI should promote the 'well-being of all sentient creatures'[74] (nice, if rather vague). AI codes also emerge in social isolation, from elite cocoons (academia, Silicon Valley) that are the very same places where the problematic models and technologies come from. And they are toothless, in that no one in those places is interested in enacting them in a way that would seriously curtail the power of the funders of those ethics programmes.

Even when we turn to some of the apparently most robust legislative responses to the problems of AI – such as the European Union's proposed AI Act – there are warning signs. This Act has promising features, categorically forbidding high-risk practices in AI, identifying those practices in areas like education, borders, crime, social credit systems, facial recognition in public spaces, law enforcement and more. But even this proposed piece of legislation (which may require some years before it comes fully into force) starts out from a celebration of the benefits that AI will bring to humanity:

> Artificial intelligence is a fast-evolving family of technologies that can contribute to a wide array of economic and societal benefits across the entire spectrum of industries and social activities. By improving prediction, optimising operations and resources allocation, and personalising digital

solutions . . . the use of artificial intelligence can provide key competitive advantages to companies and support socially and environmentally beneficial outcomes . . . [75]

If that's true, the role of regulation is limited in advance to reining in its most extreme ('high-risk') versions. Meanwhile in Brazil, a national AI strategy was published in 2021 and a legislative process is under way that is expected to lead to major legislation that encodes rights for those affected by AI systems, although the details are not yet clear.[76] The US is nowhere near drafting actual legislation, although the White House has at least published a Blueprint for an AI Bill of Rights, which frames AI governance as a civil rights issue, and there are proposals by US senators to advance legislation to rein in AI. It's worth noting that, when drafting its Blueprint, the White House consulted not only with the public and human rights groups, but with corporations like Microsoft and Palantir – exactly the sort of corporations that the document is meant to call to account.

In fact, this is the deeper pattern that underlies the dramatic statements by AI leaders in the spring of 2023 on the need to halt AI temporarily because of the 'existential risk' it poses to humanity.[77] Note, as the authors of the 'stochastic parrots' paper remark, who is talking here, and how they talk only about AI's future risks, and say nothing about resolving any of its present, well-known risks.[78]

The supposed solution is regulation, but again, the question is: by whom and on what basis? Western governments such as the UK are competing to lead the process, with Big Tech corporations guaranteed a seat at the table. Meanwhile, there is complete silence on how the views of the vast majority of the planet on this supposedly existential risk will be taken into account. Is the current rush towards AI alarmism (from those who earlier led its development) just a plea for public subsidy of the difficult and costly process of regulation? If the past is an indication, we can expect that regulation will favour corporations, not people. Too much money is at stake, and the civilising narratives tell us that the welfare of humanity hangs in the balance. As ChatGPT's creator, Sam Altman, put it, 'the most important component of wealth in the future is access to those [AGI] systems'.[79] The current furore about the future of AI, seen from the perspective of colonial history, is all about securing the continued appropriation of data resources.

To change those narratives means challenging some of the key premises of how science and technology have developed, led by the West, throughout the history of colonialism and capitalism. This, in turn, would require fundamental changes in how those technologies are implemented. If we were serious about treating AI as an existential risk and imposing social limits on its development, we would surely want to take into account the environmental costs of its production, as well as the social costs of its implementation, and to regulate it in ways that involve actual citizens rather than just high-level summits. We are a long way from this, and we will remain so, until we challenge the civilising narratives that promote a particular view of what AI is for, and whom it should serve.

Why Civilisational Stories Work

Perhaps there's an even bigger pattern here. Looking back over the stories told about data colonialism, and our brief summary earlier of the stories that evolved to justify historical colonialism, we can see that these stories, old and new, perform multiple roles.

They provide an alibi for the real processes of colonialism by telling other, more palatable stories – for example, of global commerce's necessary expansion, of the civilising mission of empire, of Big Data's enabling of connection and convenience. Believe those stories, and you are distracted from the reality of colonial resource extraction.

Furthermore, the historic idea of the West's superior science and knowledge was tied to an implied vision of the West's white racial superiority. These stories did not just enforce a hierarchy of power – they justified it and legitimated it. They worked to lock in colonised peoples to an inferior, weakened position from which change was hard. Data's civilising stories do the same: they lock citizens everywhere, except those in data's colonial elite class, into relations of continuous data extraction over which they have little if any control. They do so by telling them that this is a good deal, by the standards of the deals that capitalism has introduced into our lives. They tell them that this is the only way to assure convenient living, efficient living, connected living, enriched living.

In this way, the ruling narratives of data colonialism achieve something that the data grab alone cannot do: they capture what we might call the territory of society's imagination. They colonise our ideas of what the future of society and technology, of community and connection, might look like, crowding out alternative ideas. The result is that only one language to describe what's going on seems available: the language of connection and community offered by the likes of Meta and the manufacturers of smart devices. Connect, or else. Accept our terms and conditions, or else. Either, as Angela found, accept 'progress', or leave the game. And, in capturing our imagination, data's civilising narratives perform exactly the role that other narratives did for historical colonialism: they make it seem normal and unalterable, part of the natural order of things. Since we are only with great difficulty beginning to fully challenge and dismantle the impact of colonialism's historic narratives now – half a century or more after the collapse of its political institutions[80] – we should never underestimate the power of such narratives, including the new ones emerging around data.

The civilising narratives we have unpacked in this chapter are spreading everywhere where Big Tech is attempting to reorganise daily life. They are not just ideas. They are deeply practical ways of organising resources and people that are acquiring the force of habit. We may live with these ideas in the short term, but they lock us in for the long term, shaping our understandings of the way things naturally are, and the way things always seem to have been. We need to start challenging data colonialism's civilising myths right now. Resistance will be a long haul, but as we'll see at the end of the book, there are many starting points.

Before we get to resistance, however, we need to explore one final aspect of the emerging landscape of data colonialism: the people and institutions who are its masters and implementers.

CHAPTER 4

THE NEW COLONIAL CLASS

IT WAS KHALED'S second attempt to get on a boat to migrate to Europe, and he had already watched hundreds of videos on TikTok and Instagram from influencers who had made the same journey. He would look at the pictures of these lucky people enjoying life in Europe and talking about how, despite the challenges, the journey could be made easy if the viewers followed the influencers' recommendations.

In fact, Khaled had found the guy he was paying to take him to Europe through TikTok. Desperate for new economic opportunities that would allow him to support his aging parents in Tunisia, and to escape political instability and the advancing ravages of climate change, he ventured off.

As Khaled had learned from the videos, there was a lot of hi-tech surveillance waiting for him even before he reached the border. There were air, land and sea drones always hovering, watching silently. In fact, his first attempt to cross was foiled by a drone, although in that case it turned out to be a good thing, since the boat he was on was about to capsize when the coastguard reached them. When he finally made it to land on the second attempt, he noticed the ubiquitous towers with video cameras so powerful, someone said, that they could recognise a face from miles away, day or night. Avoiding detection was pointless, so no one even tried.

The authorities were already waiting for them, and took them to a camp. His phone was confiscated and returned only after they had downloaded and analysed all his data.

To get his meals, he had to submit to an iris scan. Some charitable organisation gave him a debit card he could use to buy some essentials, but he wondered if they were tracking his purchases. Just in case, and because he was embarrassed, he paid for some medicines he needed in cash, so they wouldn't find out about his ailments. He had also been told that the card would stop working outside the camp, so it felt like a leash to keep him there. Speaking of leashes, Khaled had heard about the ankle bracelets being used in the United States to track migrants 24/7 while immigration officials decided their fate. At least he didn't have to subject himself to such a loss of dignity, he thought.

One day, he got a WhatsApp message offering, for a price, to help him write an asylum application. The ad said that because the applications are reviewed by computers first, it was crucial to use particular words and phrases that the computer would recognise. Unfortunately, he didn't have the money needed for such an application. During his time at the camp, Khaled was grateful to have access to the free technologies offered by Big Tech to communicate with his family back home, and to access important information about his journey.

Maybe, he thought, he would be able to work for one of these companies in Europe, even as a janitor. What Khaled didn't know was that some of those same companies had lucrative contracts with governments to build the surveillance mechanisms he encountered as a migrant. According to BigTechSellsWar.com, since 2004 US government departments such as Homeland Security, Defense and Justice – whose responsibilities include not just terrorism but migration issues – have spent at least US $44.5 billion on contracts with companies including Amazon, Facebook, Google, Microsoft and Twitter.[1] Meanwhile, the EU spending on border enforcement is at an all-time high, and ties between Frontex (the EU's top funded agency) and military and security industries continue to grow.

Stories such as Khaled's, which shares features with those of millions of migrants to Europe from Africa and the Middle East over the past decade, illustrate that it is not just the *how*, but the *who* of the new data grab that matter. It matters that border security has become a multi-million-dollar business, involving coders, designers and managers within large private corporations working with increasingly intrusive technologies, often supported by powerful governments. It matters that there are large networks of smaller companies – from data brokers to app developers – that contribute to the extraction and processing of data in many sectors of contemporary societies.[2] And it matters that an individual sitting with his laptop at home, like Hoan Ton-That, founder of the US-based company Clearview AI, was able to harvest millions of faces from social media, link that data with other personalised data, and sell the resulting rich biometric dataset to law enforcement agencies in the US and elsewhere (more on this later).

But what matters even more is the larger pattern here, and its echoes of a colonial past. Historical colonialism was much more than a few acts of discovery and sporadic violence. It became a whole system for managing the world's economy and territory that evolved over centuries. And, of course, it involved people: a colonial class that did the detailed work of extracting value and managing lives, a class whose institutional power was largely, but not wholly, dismantled when anti-colonial struggles brought most formal empires to an end in the second half of the twentieth century.

Data's colonial class is only in its early decades, while the colonial bureaucracies of Spain or the British Empire had centuries to evolve. But a data colonial class[3] exists all the same, and it comprises the entire intricate networks of companies, small and large, involved in supporting the extraction of data value from every sector of the economy, working sometimes apart from governments but often in close collaboration with them.

The Social Quantification Sector

We propose one term to describe the various institutions that make up data's colonial class: the Social Quantification Sector (SQS).[4] The Social Quantification Sector is not an industry sector recognised by any labour statistics bureau, but it collects, under one umbrella, all the players involved in quantifying into data as many aspects of our lives as possible, for profit. The SQS stretches across multiple official areas of the economy: from energy to healthcare, from education to agriculture – all these are now increasingly dedicated to profiting from our personal data. So too are myriad gig economy platforms like Uber, and many specialist companies that process data without any interface with ordinary customers.

The SQS involves a range of players across the world performing a range of roles. No single corporation can accomplish the kind of large-scale extraction that we have in mind when we talk about data colonialism. Nor can any single centre of power, led from one corner of the world, orchestrate it. Indeed, as we have emphasised throughout, data colonialism is not just about the initial extraction of data: it is about a transformed social and economic order, organised around both the collection and exploitation of data, and the power that flows from that. For this reason, many types of players are needed.

In any case, it is not necessarily the size of a company that determines its impact. At its peak, the East India Company counted on a private army of 250,000 bodies (many of them Indian recruits), all managed from London by a comparatively small staff of 35 to 159 employees (depending on the period), ruling roughly 100 million people.[5] In comparison, a company like Uber has a staff of around 32,000 employees, coordinating the operations of five million drivers (not considered employees, of course), providing 7.64 billion trips to 131 million monthly active users worldwide during 2022.[6]

One other point to bear in mind at the start is that data colonialism, like historical colonialism, is a global phenomenon, although the world of today is a very different place from the world in which colonialism emerged. Any new form of colonialism must unfold within a twenty-first-century geopolitics that is very different from that of the sixteenth or eighteenth centuries, or even the twentieth century. Today's world is at least bipolar, with China's economic, technological and soft power catching up with the US and other powers like India emerging. Although many stories about Big Tech, written in and for the West, ignore China and talk only about Silicon Valley, that is a mistake we must avoid.

The origins of data's fast-expanding colonial class go back at least to the mid-1980s, when almost no one saw fully where things were heading. The first movers were credit card companies, who sought to collect data on your every purchase and from there create a profile of what sort of consumer you were, or at least sell that data on to marketers who could.[7] Various other incentive schemes, such as shopper loyalty cards, were early attempts to gather similar data. By the early 1990s, the practice was widespread enough for a prescient African-American communications researcher, Oscar Gandy, to give this new development a name: the 'panoptic sort'. Corporations were acquiring an insatiable appetite for gathering data (sorting data) outside their corporate walls. They wanted to see everything (hence 'panoptic').[8]

Their goal was to find smart ways of discriminating between their customers, seeking out the most loyal ones, and offering them incentives. And once the possibility emerged of corporations gathering data about customers beyond the mere facts of sales made to them, the search for those larger domains of everyday data began. But it is not until the late 1990s that this early vision of large-scale commercial data-gathering from society began to be reflected in management theory. We know this because in 1998, one of the US's leading management gurus, Peter Drucker, could still say that 'the more inside information top management gets, the more it will need to balance that with outside information – that does not exist yet'.[9] Drucker had sensed the direction of change: business, perhaps in response to fears of declining long-term consumption, was already looking for new ways to annex routine daily life as a source of value. No one saw anything odd about gaining such control; it was just the game to be in. As one executive put it, 'it seems to me that we have squeezed all the juice out of internal information. Maybe it's time we focussed on the whole world of external information.'[10] While expressed in the bland language of management theory, this is basically the same extractive vision that drove historical colonialism's landgrab, treating data (rather than land) as a *terra nullius*, just there for the taking.

Or at least potentially. The 'how' was still being worked out in the early 1990s. By the end of the decade (still years before the rise of Big Tech), the visionary and troubled US entrepreneur Josh Harris had figured out that the business model of the internet was based on extracting our data and then selling it back to us. Describing his experiment on surveillance, appropriately called *Quiet: We Live in Public*, he said: 'Everything is free, except the video we capture of you. That we own . . . I'm going to sell you your life back.'[11]

Fast forward nearly two decades, and a world where corporations routinely relied on a firehose of data from their customers' lives had become normal, so much so that now more outlandish predictions began to be made: for example, one marketer's prediction in 2015 that 'by 2028 half of Americans (and by 2054 nearly all Americans) will carry in their bodies device implants that communicate with retailers as they walk down the aisles'.[12]

This is the recent history of a deep social trend that sociologists call 'datafication': making the whole world into a data source for business. Some commentators compare this to a great infrastructure project worthy of comparison with the French Enlightenment's *Encyclopédie*. Well, it's certainly big, and it's becoming the basis of some sort of social infrastructure. But it's also a landgrab, or more politely a 'productive appropriation' of what was just the flow of our everyday lives.[13]

Only recently has this move become publicly controversial, following the so-called techlash triggered by the Cambridge Analytica scandal of 2018. But for today's management theorists, datafication remains just a sensible expansion of the information corporations can access, even if a profound one that some call 'a management revolution'. As the inventors of that term put it, 'each of us is now a walking data generator'.[14] But they don't mention the mechanisms of data extraction that had to be in place to make it so; they don't mention the data *grab*. Maybe that's not surprising, since, from the point of view of business, people always were in the way. As one corporate data expert put it, 'one of the major constraints to success with Big Data [is] the human factor'.[15] The group of entities we call the SQS exists to overcome that human obstacle.

What types of corporations make up the Social Quantification Sector? Let's start with an overview of the larger corporate players that make up the core of data's new colonial class.

The Big Data Harvesters

The story of data extraction is commonly told in terms of four big US corporations: GAFA – Google (now Alphabet), Amazon, Facebook (now Meta) and Apple, usually with the addition of Microsoft (making GAFAM).

To get a better sense of the different roles that GAFA members play, we can go back to the Four-X model for playing strategy video games: Explore, Expand, Exploit and Exterminate. The different phases of historical colonialism and the successive colonial powers that dominated them can be mapped onto these dimensions: Spain, the master of Explore, discovering the new territories that started the whole process; England, the master of Expand, steadily building an empire of linked territories that became the largest empire the world has seen; Holland, the master of Exploit, controlling global trading routes with fine precision; and, finally, France and the US, masters of Exterminate, experts in the elimination of resistance and the systematic colonisation of the mind.

Each of these empires applied all four X's at the same time, of course, but this broad picture also helps us roughly map the X's and the colonisers of the past onto the corporations that currently constitute the leaders of data colonialism: Amazon, the Explorer of new territories of data extraction as the basis for online retail; Google and Apple, through their dominant search engine and branded hardware, respectively, Expanding vast empires of services and infrastructures; and Facebook, perhaps the most systematic Exploiter of every possibility for extracting profit from people's time spent on its platforms. As to Exterminate, it is harder to map this more brutal dimension onto a single corporation or a single practice, but at the same time, as we have outlined, many of the practices of data colonialism do result in direct physical violence or in other forms of indirect or symbolic violence, including loss of opportunity, deteriorating welfare, or the eradication of alternative ways of thinking or being in the world.

The point is that, merely two to three decades from its inception, data colonialism represents an evolving landscape. The way these corporations have succeeded in mastering the larger economic opportunities of data extraction deserves a closer look.

Starting in the mid-1990s, Amazon led the way in opening up the territory of e-commerce and online retail business. It now has a market capitalisation of US $1.28 trillion.[16] The company generates a large majority of its income (72 per cent) from its online store platform, advertising, and services to third-party sellers.[17]

But Amazon is much more than just a store. Data from tracking customer habits is key to its income, and Amazon has many ways of grabbing our data. In the home sector, it has built the world's leading digital personal assistant, Alexa, which is the market leader in the US (think of all the Echo devices, including Loop, Ring, Frames, Smart Oven, and the many third-party Smart devices that can be controlled through Alexa, all while capturing data about our habits).[18] Amazon would love to take a look inside your phone, which is why it is piloting a program that offers some 'lucky' users US $2 a month to let the company monitor their mobile devices[19] (you can rest assured the value that Amazon can derive from this data is much more than a couple of dollars). Crucially also, Amazon is the world leader in the cloud computing business, with Amazon Web Services (AWS) capturing 32 per cent of the world market,[20] making up 16 per cent of its total income. [21] Even the US Central Intelligence Agency (the CIA) is an AWS client, which speaks to the level of trust in its security protocols.

Google, now part of Alphabet, has a market capitalisation of US $1.57 trillion. Since 2010, Google has consistently led the world's search engine market, with a share of 92 per cent in June 2022 across all platforms, including mobile.[22] By this date, advertising comprised 81 per cent of Alphabet's income, with the bulk of it (58 per cent) coming from advertising linked to search. But Alphabet has many other major business lines that grab data: it owns You-Tube, dominates geographical information (Google Maps), and has major investments in education, health, home personal assistants (Google Home), and more generally AI services. It also has a small but growing Cloud business.

Alphabet is increasingly focussing, for the long-term, less on the data about us it can sell to advertisers and more on using its own vast stores of integrated data to offer an ever-more personalised service to its users, what CEO Sundar Pichai has called 'each user['s] own individual Google'.[23] Although some believe its main search engine and advertising business will be damaged by the revived competition from the ChatGPT-enhanced search engine Bing that Microsoft owns, this may underestimate Google's strategy of expanding its domination across multiple knowledge services basic to everyday life.

Facebook (now part of Meta) is the member of the Social Quantification Sector most in the public eye, in part because of a series of eye-watering scandals, from Cambridge Analytica in 2018 to the revelations of whistle-blower Frances Haugen in 2021, spotlighting how ruthlessly Facebook and the other parts of Meta (Instagram, WhatsApp) have been willing to exploit their privileged access to the so-called social graph of Meta's users (the mathematical patterns that capture the shape of its users' social networks). Whatever its troubles, the Facebook platform is the world's largest social media platform, with almost three billion users by April 2023 and a market capitalisation of US $720 billion.[24]

Meta is, however, substantially less profitable than Alphabet and Amazon, and its income is almost entirely dependent on advertising.[25] That advertising profit took a major hit in 2022 when Apple changed iPhone's terms of operation to force third-party advertisers to give its users the explicit right to opt out of being tracked, although by early 2023, there were signs Facebook was developing ways round this by utilising its own AI to produce a bespoke ad-targeting service based on its own exclusive data.[26] Meta has also recently become the first of GAFAM to face major global competition, from TikTok. And its plans to convince us all to migrate into its virtual Metaverse don't seem to be going very well, so far.[27] But it would be foolish to write off Meta just yet.

Then there's Apple, with a market capitalisation of US $2.9 trillion. Apple is an outlier in the data colonial class, in part because of its strong public stand against generalised data collection. But this stand is grounded in its own highly distinctive business model.

Instead of generating predictive data about its customers, Apple makes money by locking customers into its very own data colony: Apple's universe of hardware devices whose operation and maintenance Apple exclusively controls. The strategy has paid off substantially: by early 2022, it had 62 per cent of the premium phone market (phones priced at more than US $400).[28] This generates its own guaranteed streams of revenue, which flow not just from devices (Apple remains by far the most profitable phone manufacturer in the world),[29] but also increasingly from what Apple calls 'services', which include the commission Apple charges to those who develop apps for the iPhone.[30]

Apple's recent use of its huge territorial power in the smartphone market just mentioned has laid the ground for Apple's own advertising business to grow unimpeded – based on the data Apple exclusively collects on its phone users, as well as the exclusive content that Apple produces, such as Apple TV and indeed its App Store![31] That data is used, for example, to target ads, and to recommend apps and games on the Apple Store, just as Google does on its Play Store. This has even led to a lawsuit, claiming that the company collects data without users' awareness, even when they select the 'do not track' options on their iPhones.[32] Apple's relations with China are also notable. Not only are 95 per cent of iPhones still produced in China, but Apple is also, currently, the most profitable Big Tech company in China.[33]

China's Parallel Data Empire

The landscape of the world's large-scale data harvesters is highly concentrated,[34] but so far we have only considered US corporations. China has built its own parallel empire of data extraction, the so-called BATX: Baidu, Alibaba, Tencent and Xiaomi. While these companies operate mainly (but not exclusively) in Asia, the emergence of TikTok (owned by ByteDance, valued at approximately US $220 billion in March 2023)[35] suggests a shift to a more direct competition between US and China, the contours of which are not yet clear, given US threats of retaliation.[36]

Though not so well known in the West, the members of BATX match GAFA in range of market power, at least in China. Amazon's Chinese counterpart in online retail sales is *Alibaba*. Recent market turbulence reduced its market capitalisation to US $236 billion. Like Amazon, the bulk of its income comes from consumer retail.[37] Unlike Amazon, however, it long ago moved into personal finance, creating China's most popular payment app, Alipay, which 93 per cent of Chinese online buyers say they use.[38] It is worth noting that Alibaba has increasingly tense relations with the Chinese government, which have affected its standing in global stock markets.

Google's Chinese counterpart *Baidu* is also a powerful search engine for the Chinese market, launched in 2000. Like Alphabet, it has a large investment in AI, but it has also branched out into ride-hailing services, launching a driverless ride-hailing service in mid-2022.[39] It also has its own Tiangong platform for smart IoT and the DuerOS voice assistant for the Chinese market.[40] Its market capitalisation, however, is much lower than Alphabet's (US $50.8 billion).

Facebook's closest counterpart in China is *Tencent*, which in 2011 developed the super-app WeChat discussed in Chapter Two. Tencent (market capitalisation of US $444 billion) is certainly of comparable size to Meta. But like Alibaba, Tencent moved into personal finance early, and its payment app (WeChat Pay) is almost as popular among Chinese customers as Alipay.[41] Tencent is much less dependent on advertising income than Meta.[42] It is also now a key portal to access government services, which Facebook is not. Unsurprisingly, Tencent's relations with the Chinese government are also becoming more troubled, which has begun to affect its market capitalisation too.[43]

Meanwhile Apple has something of a counterpart in *Xiaomi*, although it is the least known of the BATX and very much smaller in terms of market capitalisation (US $35.7 billion). Yet it is the world's second largest shipper of smartphones after Samsung.[44] It is also a leading player in China's Smart Home device market, paralleling Amazon, and has a popular health-tracker called Mi Band that forms part of its Mi Fit platform.

The influence of Chinese Big Tech extends beyond the platforms that operate in China to its involvement in digital infrastructure building, for example across much of Africa. In this context, it's worth mentioning the little-known Chinese phone hardware company Transsion (market capitalisation US $15.7 billion as of June 2023[45]). Transsion is the largest phone supplier to the African continent. With production across multiple sites (India and various African countries) it has specialised in phones whose hardware, with multiple pre-set apps, does not require huge memory, and so fits better with the technology access issues that affect Africa. Transsion is becoming an increasingly important player in 'the datafication of everyday life' in Africa, operating under the radar of the US or even the Chinese big players.[46]

The New Colonial Vision Distilled

Each corporation in the Social Quantification Sector has a particular vision of why the data grab makes sense for them and for the world. Consider Microsoft, an often-neglected player in the recent story of Big Tech, even though it has a market capitalisation of US $2.5 trillion. Close to 1.5 billion people and businesses use Microsoft's word-processing products and services worldwide,[47] even if it is much harder to put an exact number on the hundreds of millions who use its Office 365 product. It is dominant in business networking, having bought LinkedIn in 2016 and the software developer platform GitHub in 2018. The company also plays a major role in the data infrastructure behind the most advanced techniques for personalised marketing through its ownership of Xandr, a leading data solutions provider for the marketing industry. And it is a major rival to Amazon in cloud services (with 23 per cent of the global market for cloud infrastructure),[48] playing an important role in health data and EdTech. As one of the key investors in Open AI, the developer of ChatGPT, Microsoft has recently been in a strong position to exploit AGI to augment its own search engine, Bing.

The company's current chief executive, Satya Nadella, is one of the most eloquent representatives of data's new colonial class. He is not, however, the standard aggressive corporate type. Nadella is a self-reflexive product of earlier generations' critiques of capitalism (he talks openly about his Marxist parents in India). And his management philosophy does sound inclusive, judging at least by his 2017 book *Hit Refresh*.

Nadella wanted from early on to produce 'software that would change the world'. He started in a background role, working on the technical infrastructure of Microsoft's cloud system. From his perspective, a flexible cloud that supports business's growing needs for data processing is a way of 'democratizing and personalizing technology'.[49]

But the core of Nadella's vision should certainly make us pause. That vision is to create a world where data capture happens seamlessly in the background without obstruction. Inspired by novelist Neal Stephenson's concept of the Metaverse, Nadella talks of 'replacing reality' 'with a digital imitation'. These ideas, he adds, 'are now within sight'. Microsoft's detailed ambitions for the Metaverse have recently been announced, and they rival Meta's in scope: those ambitions have driven its recent expansion in the games market (the controversial acquisition of game developer Activision Blizzard for almost US $69 billion).

But perhaps it is in Nadella's description of Microsoft's little-loved digital personal assistant Cortana that his thinking comes across most clearly. Cortana, he writes, is 'an agent that knows you deeply. It will know your context, your family, your work. It will also know the world. It will be unbounded. And it will get smarter the more it is used. It will learn from its interactions with all of your apps as well as from your documents and emails.' That's a vision of total surveillance, unless you believe that the data gathered stays entirely under your control, which does rather stretch our credulity. And even if you do, what happens when similar business models are applied by other corporations?

The Wider Colonial Class

In data colonialism, each platform or app is a potential data territory where, under the developer's sole sovereignty, data of certain sorts can be gathered, actions of certain types can be induced, and relevant predictions about the user can be made. The resulting information and profiling can be held for the benefit of the developer, or sold for others to use. The only catch is that the platform or app must find users to generate the data, for which they depend on app stores controlled by Apple, Google Samsung, or other SQS big players (these companies extract a substantial cut from any sales made through their stores).

There is no limit to the number of platforms and apps that can be built on top of our online activities. It is getting hard to do anything in life – buy a ticket, find somewhere to stay, find a recipe – without using a platform or app. They are becoming an infrastructure for daily life. Businesses can go on adding endless layers to this infrastructure, as long as people can be induced to use them.[50] In principle, there is no limit to the extraction opportunities because, as explained in Chapter Two, data territories have no limits in physical space.

True, not all apps extract data from us for their own benefit. We certainly *hope* our local bus and train apps don't, since they relay to us information about the public service we already pay for, although perhaps we should know better.[51] The default position is to extract data from whatever customers do, no matter how trivial.

Needless to say, large numbers of coders, designers, managers and marketers are required to keep the system going. Each is a member of data's colonial class, the network of people who sustain and expand today's data grab. The result is an emerging bureaucracy that, in time, may parallel the bureaucracy of colonial administration that ran modern empires. Yes, given the market ownership of all the technologies that implement data colonialism, we would expect this administrative power to be more distributed than the bureaucracies of historical empires which – as for example in British East Africa – had a very close association with state power. But the core function, to manage economic opportunities and control populations so as to ensure the optimisation of those opportunities, is basically the same.

The Data Scientist and Today's New Colonial Mission

All of this shows that there is obviously a human side to data colonialism, just as there was to historical colonialism: individuals attracted by the uncertain but maybe boundless riches to be gained from new territories, individuals working together while believing in the same civilising narratives. Management theorists leading the Big Data revolution are more than happy to talk up the excitement of this work, calling the data scientist 'the sexiest job of the twenty-first century', or as another data scientist said: 'the future belongs to the data scientist'.[52]

In one way, such claims cannot fail to be right. The controller of a data territory really does rule the future of data extraction in that space. And if that space is increasingly interwoven with our daily lives, or with the information flows of a whole corporation, the future-managing power of the data scientist can only grow. As one of them put it: 'When the company sees what can be created with data . . . you'll see data products appearing everywhere. That's how you know when you've won.'[53] The horizon of data extraction stretches out like the territory that colonial administrators once surveyed: there to be captured and dominated.

This vision of control through data – through expanding territories of data extraction – is intoxicating. It is fuelling a vision of extended data grabs by 'intelligent corporations'[54] that seek to control not just individual supply chains, but entire production and consumption cycles. International NGOs like the World Bank and the United Nations, far from offering values that challenge the data grab, have been enthusiastic partners in this. Their own development projects routinely impose on Global South populations heavy obligations to generate data. These are the same populations whose economic and social disadvantage were what 'development' was meant to alleviate in the first place. But that always seems to be the case: some pay for 'development' more than others. If data power is a way of gaining social power through holding information, then it is also a new way for organisations based in the Global North to expand their power in the Global South.[55]

This global inequality is reflected also, as in historical colonialism's governing class, in the power relations that shape who is part of the new Big Tech elites. Although data science offers a whole new technique of management power for data's colonial class, the opportunities to join that class are not equally distributed, any more than they were to join historical colonialism's bureaucracy. If someone is white, their chances of entering data's colonial class are much higher than if they are not: just 7 per cent and 8 per cent of jobs in US high-tech companies are held by Black and Latinx workers respectively, almost twice as bad as the representation in US industry generally.[56] If someone is male, their chances are also much higher: in 2021 only 29 per cent of employees in tech companies worldwide were women.[57] The emergence of data science as a profession and the costs required to train a data scientist (economic costs to the economy, and financial costs privately borne by individuals and families) have ensured that the opportunities to enter data's new colonial class are not open to all. And this is not just a matter of the general inequalities that shape Western societies' distribution of work. In part, this is also because of the continuities between today's data sciences and the history of colonial science itself that we discussed in Chapter One. Whatever the cause, the 'future' that supposedly belongs to the data scientist is, to paraphrase William Gibson, not evenly distributed.

Who's Watching When You Drive?

As we've suggested, data extraction is not just something happening in social media or even retail marketing, but across all areas of life where a computer has been inserted. As a result, the SQS is much larger than might first appear.

To illustrate this, take car-driving and the insurance that goes with it, for many people a basic aspect of everyday life, which is being transformed into a data territory (we'll stick to cars we drive; self-driven cars are data territories from the start).

Any new car today is likely to be Smart, that is, data-extracting. Cars have essentially become computers on wheels, and those computers collect a lot of data (it's very hard to buy a new car today that doesn't transmit data). We might appreciate the convenience of car systems that can help us drive or even drive on our behalf, protecting us from potential accidents by detecting threats and reacting to them faster than we can, or even just reminding us that we are due for an oil change. But cars these days collect – and more importantly, report – lots of information about us: how many passengers are travelling, whether their seatbelts are activated, whether the doors are unlocked, what the cabin temperature is set at and whether the windows are up or down, and even what music or entertainment programming we are listening to. Smart cars also record extensive trip data: location, speed, destination, when the brakes are applied and when the car accelerates, whether the headlights are on, and so on. All of this information may say something about us as drivers and people, and all of this information is transmitted wirelessly not to us for our edification, but to business.

As a result, cars are not just a mode of transport, but a large data market in their own right, which analysts predict will be worth anything between US $300 and $800 billion by the end of the decade. Data extraction in the car sector involves basically three types of corporate players who manage the data territory that cars now represent.[58]

First, there are the data harvesters, the car manufacturers who install devices for extracting continuous data into the computer-based systems that operate the car. Those systems 'capture' data like any other computer has for the past three decades. In the industry, carmakers are known as OEMs (original equipment manufacturers).

Data is now the focus of a number of economic alliances between OEMs and third parties. There are suppliers of navigation systems and internet-enabled infotainment systems such as Apple CarPlay and Android Auto, and, if the car has a built-in phone connection, telecommunications operators. Then there are the specialist manufacturers who make installable tracking devices to collect even more data, the so-called 'aftermarket' data gathered when, after the car is sold, the owner agrees to have an additional data-tracking device installed. The Electronic Logging Devices (ELD) for trucks discussed in Chapter Two are an example of an 'aftermarket' device. Finally, there are the Big Tech suppliers of voice-activated search functions: yes, you guessed it – Google, Amazon and Apple. Their data extraction is so significant that one car data specialist has called the car 'a browser's last frontier'.[59]

What happens to all this data? That's where a second member of the new colonial class steps in: the data aggregator. A whole specialist sector now exists to process the mountains of data generated by our cars, other than that already siphoned off by Big Tech providers. In the car industry these specialist data aggregators are called 'vehicle data hubs'.

Big players such as LexisNexis (based in the US) and Otonomo (based in Israel) are in this market. They perform a role which has equivalents in many other sectors, such as, in marketing, large companies like Experian, who aggregate multiple sources of data and

process it into a usable form. In the background, Alphabet is playing an increasingly important role in setting the standards for
smoother data extraction in the car sector.[60] Much of the data gets
anonymised for broader industry analysis, where individual identities are irrelevant. Similarly in the car sector, vehicle data hubs can
generate anonymised, aggregated data for general use. But if the
individual inputs to such larger datasets were ever combined with
other data so as to identify an individual, there are third parties
interested in buying such data for sure (for example, advertisers
might want to know your preferred cabin temperature before they
try to sell you a hot or cold beverage).

But data hubs also interface with a crucial third player in the
car-data market who absolutely does *not* want anonymised data.
This is the insurer. Insurance contracts are increasingly being
turned inside out through data processing. The new 'usage-based'
policies (sometimes called 'connected' policies) no longer protect
us against one-off unknowable risks which might affect anyone at
some point during their driving career. They are products tailored
for particular policy holders and continually refined in the light of
the data that is gathered directly from a particular driver's car.
Instead of having to spread general risks across a pool of drivers,
insurers now have the option to take on much more narrowly
defined risks tailored to precisely the sort of driver they observe
each of us to be.[61]

Insurance contracts are now, in other words, surveillance contracts, although that is not how they are marketed. A blog by The
Floow, a company acquired by vehicle data hub Otonomo to run
such a policy, is quite explicit. It describes how it installs a device in
the car to 'monitor [drivers'] habits and behaviors'. Note the word
'habits', which suggests there's plenty of larger-scale analysis done
over and above the moment-to-moment tracking.[62] It is not for
nothing that one such policy, marketed by Progressive Insurance in
the US, is named Snapshot.

The colonial dimensions of this new type of car insurance become even clearer when we look at how the same business opportunity plays out in the Global South. 'Personalised' car insurance is also proving attractive to insurance companies in Latin America, Africa and South-East Asia. Why? Because, as a group of Swiss political scientists recently found, it provides a great new way for insurers to expand their clientele to populations whose risk factors, in insurers' perceptions at least, might otherwise be unattractive. This is accomplished by expanding the range of risk-relevant data insurers extract, not just from the car (like GPS location data) but also from the driver's personal details and social media accounts.

The researchers interviewed insurance professionals and found plenty of evidence of seemingly bland categories, like 'personal address', being used to hide implicit judgements about, for example, race. Your address, after all, is rarely unconnected to your race, no matter where you live. Given that, it's hard to have confidence that other categories such as 'likely to fail' will be used without subtle reference to racial factors. What we can see here, in a particular racialised form, is how the data grab is opening new types of markets where service providers no longer have to socialise risks, distributing them across a wide population. As the CEO of a Kenyan-based insurance platform put it: 'we thought that it's not fair to charge everyone the same 5 per cent because some people are better drivers than others. So now our regression model comes up with this percentage' (translation: if the data extracted by the company suggests you are likely to be a bad driver, the algorithm will also calculate how much more you will pay).[63]

The jury is still out on whether this trend in car data extraction will go on expanding, at least in the Global North. Some experts think car manufacturers, picking up on drivers' squeamishness, will start resisting insurers' claims for a complete data stream, with similar tensions emerging between carmakers and Big Tech corporations like Google.[64] But, if you can afford a Tesla car and are interested in Tesla's insurance offer, you won't have a choice. Tesla cars extract large amounts of data as they are driven,[65] and the only car insurance Tesla offers is tailored to the data extracted about owners and their cars. Indeed, such data, from an industry perspective, enables a new sort of 'value-added service': Tesla doesn't just collect data about you; it's happy to sell you back your data, for an additional fee, as interesting insights![66]

Economists tend to see this new form of data extraction as just what's needed to correct the information deficits that have held back the car insurance market until now.[67] But that is to miss the wider extractivist transformation that is under way.

Serving the Algorithmic State

We have covered the range of commercial actors involved in the Social Quantification Sector. But where does the state and national government fit into this? There are probably just as many areas of data extractivism in the public sector as there are in the private sector. Indeed, the two are becoming increasingly intertwined.

It is generally assumed that the government must store some data about us (securely, we hope). But when the state uses data to target specific populations without transparency or accountability, things get a lot messier. Consider the region of Catalonia, in northern Spain, where it has been revealed that the government maintains an extensive 'preventive' techno-social system to monitor the Muslim population. A few misinterpreted Facebook posts, along with signs of involvement in anti-Islamophobic groups, can be enough to deny a citizenship application and expel the applicant. And while the system relies on more than just digital methods of surveillance, there is no question that technology plays a big part in the Catalan Protocol for Prevention, Detection and Intervention in Processes of Violent Extremism (PRODERAE in Spanish).[68]

This kind of surveillance extends to a nation's borders. Remember Khaled's story from the beginning of this chapter? States are increasingly relying on digital data to control the movement of migrants through what they call 'smart borders'. Unsurprisingly, the datafication of the border is not resulting in more humane treatment of migrants. According to United Nations High Commissioner on Refugees, at the end of 2022 there were 110 million refugees in the world fleeing poverty, violence, human rights abuses or environmental destruction.[69] For many of them, the attempt to cross a border entails an encounter with a multitude of technologies for extracting data, as we saw with Khaled.

Smart Borders – we came across that word 'Smart' in Chapter Three with the Smart Home, but this usage is unquestionably more sinister – are often more lethal than 'low-tech' borders. By increasing the pervasiveness of surveillance, they push migrants to attempt crossings at more dangerous (but not as highly monitored) locations, which has resulted in higher death rates.[70] In addition, Smart Borders are more prone to mismanagement, because the tools that make them smart are generally produced by the private sector, making regulation and accountability more difficult.

Many technologies are involved in Smart Borders: automated systems to review visa applications; autonomous surveillance technologies such as drones to monitor air, sea and land; intelligent video systems that can track and identify humans; extensive biometrics (including facial recognition, fingerprints and iris scans) to control migrants at border crossings; the use of ankle bracelets or other location-tracking devices; and an extensive framework for extracting, analysing and cross-referencing data from government records, medical and education records, phone location data, social media, and other sources of personal information.

Behind these vast efforts overseen by states are many SQS companies, including household names like Amazon and Microsoft, whose cloud services are relied upon to store and process the data extracted. But the sector also involves other less well-known corporations who play a crucial role in processing data at the largest scale, such as Palantir, Salesforce and LexisNexis. These companies receive millions of dollars in government contracts, in complex arrangements, which leave it uncertain whether data collected for one purpose then gets used for another.

The colonial echoes of smart borders are striking. Data technologies of some sort have always played a role in the control of colonial borders. Indeed, that was why many techniques – from passports to fingerprints to immunisation cards – were invented. And while it is not easy for some humans to bridge national divides, it is easy for technologies to do so; solutions applied at the border to track and surveil populations eventually find their way to the rest of the country, or to other countries. Think of the multiple uses of facial recognition technology for everyday policing in autocracies like Iran, where they were re-deployed to monitor women's face-coverings in September 2022.[71] Borders have always been places where the nation-state's authority is violently enforced, within the wider logics of colonialism. The connected power and unlimited scope of today's technologies of data extraction have deepened the asymmetries involved.

Palantir's Data Territory

Controversial Silicon Valley entrepreneur Peter Thiel once remarked: 'My generation was promised colonies on the moon: instead we got Facebook'.[72] Although Thiel was an early investor in Facebook, he soon founded his own operation. We don't mean the payment platform PayPal (which he also founded),[73] but rather his role in setting up Palantir Technologies, a company which has a major role in supporting the contemporary state's border operations. This will be our case study of how data's colonial class is serving the state's wider goals of data management and territorial management.

Palantir is a US company quoted on the New York Stock Exchange (market capitalisation, US $34.5 billion). It was founded in 2003 and remains one of the most secretive Big Tech companies.

It operates a bit like the data aggregators we found in the car sector or marketing: it doesn't collect data itself but provides crucial data integration services to a range of external clients. Palantir's clients are mainly in the security, defence and intelligence sectors, but the company also provides services to corporations and organisations in healthcare, drug safety, and other areas. Its role in providing the infrastructure of the UK National Health Service's data processing has become increasingly controversial.[74] But it is in relation to core government services that Palantir's greatest importance lies.

Governments gather lots of data when people cross borders. They also have access, directly or through their own indirect powers, to vast amounts of other data. But how to link these data sources and make sense of them? And how to do this systematically, continuously and in real time, to target an emerging threat supposedly posed by an individual crossing a border right now? This is where Palantir's services come in. Palantir does a lot more than aggregate heaps of data into larger datasets. Without getting too technical – and the details are *very* technical – Palantir works on three interconnected levels of increasing significance. To get a sense of how they work, let's imagine a government client for whom Palantir is consulting.

First, Palantir organises the multiple messy datasets that the government may have relevant to an area. In other words, it gets one database to talk to another database (connecting the Student and Exchange Visitor Information System to the Alien Criminal Response Information Management System, for instance). Government itself may have no idea how to integrate them; indeed, they may think it impossible to get those datasets to talk to each other. But Palantir tidies up the inconsistencies, labels everything systematically, and generally turns that pile of zeros and ones into one very large and functioning dataset. Incidentally that is exactly the advantage that some National Health Service (NHS) managers in the UK see in Palantir's core software, Foundry.[75]

Then Palantir goes deeper, looking for the underlying patterns that may be hidden in that vast terrain of the now organised data. Techies call this an 'ontology'. A more familiar term for what Palantir creates at this stage is 'metadata', generating ever more subtle ways of labelling and ordering data to reveal deeper patterns across billions of data points.

The third level is where the real action is: that's when the deep patterns Palantir has discovered get mobilised to generate predictions, revealing this or that complex data pattern to be associated at high probability with a specific *future* offence or risk. Armed with this prediction, Palantir supports the government to take pre-emptive action in individual cases, cutting off a person or a line of activity before that predicted outcome ever materialises.

In this way, Palantir has become an active part of how governments actually govern. Much more than background support for the undeniably complex tasks of contemporary government, Palantir's systems enable decisions to be generated from vast unstructured domains of data: data from which decisions could never otherwise have been generated. Palantir finds patterns across multiple vast databases that have never before talked to each other. And it creates new targets ('data objects' is the technical term) that have never before been identified as targets; they only emerge as objects from Palantir's multi-layered processing. Once such targets have been isolated, decision-making and planning becomes reorganised around them.[76]

For example, Palantir provided a tool to the US agency for Immigration and Customs Enforcement (ICE) that made it possible, across multiple databases, to identify undocumented individuals relevant to potential investigations. The result was to target, arrest and deport family members of children who crossed the US border alone and in need of sponsors, effectively preventing these children from joining their relatives in the US. While Palantir claimed its software was never used in deportation cases, at least 443 individuals were deported by ICE as a result of this particular operation.[77]

Palantir literally reconstructs the world on which government acts. In technical terms, according to one US Securities and Exchange Commission (SEC) filing, it fulfils its clients' needs for 'generalizable platforms for modelling the world and making decisions'. But these are not one-off models that, once used, are put to one side. They are an emerging and permanent new infrastructure of government decision-making. Where? In the US certainly, but Palantir's activities in Europe have proved harder to track down, although it is known to be active in relation to security services in Denmark, France, the Netherlands and UK.[78] Major crises are Palantir's friend: it advised the Greek and Dutch governments during the Covid-19 pandemic and the UK government in handling the Ukraine refugee crisis.[79] While Palantir might blandly label this 'Data as a Service', echoing the better-known business product of 'Software as a Service', its real ambition was more clearly expressed in its SEC filing when it said it aimed to 'become the default operating system for data across the US government'.

A recent exhaustive study of the language of Palantir's registered patents uncovered something even more concerning. An ambition to change what counts as knowledge for those in power, but also, in time, for those on whom that power is imposed. As the study's authors put it: 'Palantir imagines a world where its platform might serve as a "shadow" universal knowledge graph for governments, industries, and organizations'. In other words, Palantir holds up a mirror to the social world and persuades powerful organisations and national governments to recognise their territory in it,[80] a mirror that didn't exist before Palantir came on the scene.

That mission is reflected in the details (only starting to emerge as we write) about the role that Palantir is playing in supporting the Ukrainian army against Russian invaders. By bringing together data from commercial satellites and secret intelligence sources, combined with on-the-ground sensors, Palantir and the Ukrainian army have built a system that allows commanders to remove the lack of situational awareness commonly known as the 'fog-of-war'. This system is all the more powerful because the data can be processed by powerful AI agents halfway around the world, and then fed back to the battlefield in real time (using Musk's Starlink internet satellites). According to experts, the advantage this system provides is similar to the advantage nuclear weapons would give an army over another one that doesn't have them, and it is changing the way wars will be fought for decades.[81]

Palantir is just one prominent example of how the twin logics of corporate and governmental data extraction work together in the era of data colonialism. The upper echelons of the data colonial class collaborate to transform what counts as 'actionable' knowledge in our societies. For the Ukrainians, this knowledge appears to be giving them an edge over their enemies (although we might worry about Palantir selling their technology to an army we *don't* support).[82] For Khaled, the consequences of this knowledge are felt in the constraints under which he has to live.

At this point, barely twenty years into Palantir's history, we cannot tell where this will lead. It took the East India Company a century and a half after its founding in 1599 before it became the military quasi-state that we now remember.[83] We should keep in mind where Palantir's corporate name came from: 'Palantirs' were the all-seeing stones in Tolkien's *Lord of the Rings* that saw into events, both past and future. The question is still: who has control over such power, and to what ends?

The Emerging Global Data State?

Palantir is only a spectacular version of a much broader transformation. Right across the global economy, Big Tech companies such as Microsoft and Google are offering themselves as the information infrastructure for business and governments. Without such large-scale services for integrating data and storing it in the cloud, they claim, businesses would be unable to manage their relations with the world (Palantir itself uses Google Cloud).

Far from being a recent development, such links are part of a much longer, and largely hidden, history of state-corporate collaboration. In the US, Alphabet (Google's owner) from early on benefitted from close links with the US Department of Defense and, as economist Mariana Mazzucato has shown, even the famously independent Apple relied on research done by government-funded scientists for the technical innovations that underlay the iPod, iPhone and iPad. Meanwhile, it is well known that it was DARPA (the research arm of the Department of Defense) whose funding started the internet itself.[84] No surprise therefore that in December 2022, Alphabet was selected by the Pentagon to deliver a US $9 billion cloud computing contract, alongside Amazon, Microsoft and Oracle.[85] And no surprise that DARPA funded the development of voice-activated interfaces, which is the core innovation of the voice intelligence industry.[86]

Against this background, commentators in the West have no right to be surprised, in turn, that the Chinese government strongly supported its key platforms in their early years.[87] Recent conflict between the Chinese government and platforms such as Alibaba, Tencent and ride-hailing platform Didi (which in June 2021 was pulled from China's app store days before its listing on the New York Stock Exchange) has demonstrated, in reverse, the significance of the early political backing necessary to gather and process data in a society such as China. The Fourteenth Five-Year Plan, released in April 2021 by the Chinese Communist Party, names data for the first time as one of the nation's core assets, suggesting that much greater state control is on its way.[88]

What is most striking about Big Tech's unfolding in China, however, compared to the West, is the presence from the start of a larger strategic vision that integrates 'consumer, corporate and government data' into one national asset.[89] It's worth looking at China's policies for the development of its 'internet plus', Artificial Intelligence and Social Credit System, which fortunately have been translated into English.

The policy document authorising the planning of the social credit system lays out the core of this vision. The Chinese government does not use terms like 'empowerment' or market 'freedom', so common in Silicon Valley. Instead, it talks more directly of 'a market improvement of the social and economic order'.[90] Yes, order, which is fundamentally a social goal, not obviously aimed at freedom. Yet the term 'order' also describes rather well many of the social transformations through data that we've unpacked in previous chapters.

At last, we see the wider logic that drives the expansion of data's seemingly disparate colonial class, wherever it is, and whether employed by corporations or governments. We are in the presence here of an extraordinary ambition: the ambition to know the world in new and enduring ways, and on that basis to exercise territorial power with an intensity not previously known in history. If corporations, as the main seizers of data territory, are the initial beneficiary, we can be sure that governments will not want to be far behind.

It is easy to be distracted by examples of authoritarian government overreach. We have seen much hand-wringing in Western media about China's Social Credit System, or about the Indian government's Aadhaar programme that requires biometric ID for even the most basic dealings with the state, resulting in what some call a 'welfare-industrial complex', that is, an alliance of private corporations working with the state to extract commercial opportunities from both welfare system and conventional industrial development.[91] But all governments, including democratic ones, are benefitting from data-driven schemes across a wide range of their operations.

What they are benefitting from, very often, are the seizures of data territory that *corporate* techniques everywhere make possible: that, after all, was the key message of the Snowden revelations in 2013. So let's not fall for the story of good corporation/bad government, let alone the idea of good democratic government using tech in a good way/bad authoritarian government using tech in a bad way. The very nature of government is being transformed everywhere by the new form of control and knowledge that data territories make possible through a 'public-private surveillance partnership', in which old governing classes work closely with data's new colonial class to re-order power relations.[92]

There is an important continuity here with the histories of both colonialism and industrial capitalism. Government has always depended on the gathering of information about whatever territory it claims to govern, but the state's ability to gather information has varied hugely over the centuries. Government's data-gathering accelerated most in the nineteenth century, when colonial powers honed new surveillance techniques in their colonies such as the fingerprint and, in order to manage a home population suffering from the disruptions of early industrialisation, refined statistical techniques.

Today governments are increasingly trying to integrate the commercial techniques of data analysis and algorithmic processing into their everyday decision-making. In many cases, they rely on the infrastructure provided by Big Tech (it is reported that 6,500 US government agencies are already using Amazon's cloud services[93]). Latin America is one region where such processes are expanding, as governments with few resources look for quick routes to solve intractable social problems. In Argentina, for example, the government recently signed a contract with Microsoft to use AI to create a 'Technology Platform for Social Intervention' that predicts teenage pregnancy among lower-class women.[94] Horus is another Latin American social welfare project, prominent in Argentina and Brazil, that seeks to cut welfare waste through more intensive profiling of welfare recipients, once more driven by the AI services of Big Tech companies like Microsoft.[95]

Meanwhile, as we mentioned earlier, the government of India has built the Aadhaar identification system as the infrastructure for social welfare, and much else, to which 89 per cent of the population were signed up by mid-2019.[96] Its early justification was to give many poor citizens a presence in government databases, which would make it easier for them to file claims under India's vast welfare system. But it has expanded into a much more comprehensive way of underpinning India's civic and economic activity, with some similarities to China's system. From the start, it was based on biometric data and a public-private partnership, generating serious privacy and surveillance concerns.[97] Data in India, as in China, is proving an effective focus for state-driven nationalism.[98]

By contrast, as the example of Estonia and its X-Road project demonstrates, it is extremely rare that a national government decides to run its services on the basis of a perfectly conceivable alternative: that data flows should remain under the sole control of the *citizens* whose data is being used.[99] So far there is no evidence that Estonia's example is being copied by governments elsewhere!

Data's Lone Adventurers

The story of Big Tech, when seen the right way up, is really the story of data colonialism. The main characters are usually large entities (corporations, governments). But as with historical colonialism, individual adventurers play an important role too. They too can be conquerors or conquistadors, and engage in what one writer has called 'surveillance frontierism'.[100]

A curious example of this occurred when in May 2022 two powerful men shook hands on a deal that they hoped could potentially bring them great benefits. One was Brazil's then-president Jair

Bolsonaro. The other was Elon Musk, president of Tesla and SpaceX, and now owner of Twitter. The handshake, sealed in a luxurious resort in Porto Feliz, signalled an intention for Musk's Starlight satellites to bring the internet to 19,000 isolated rural schools, and to monitor the environment in the Amazon rainforests.. It is doubtful now the agreement will go forward under the administration of the new president of Brazil, Lula da Silva, but it casts a retrospective light on the ambitions of more than one type of adventurer. For the real purpose of this proposed deal, we might surmise, wasn't doing good to schoolchildren or the environment, but data extraction, plain and simple (meanwhile, Musk's Starlink system is already being used by illegal miners in the Amazon).[101]

We met another adventurer earlier in the chapter: Australian Hoan Ton-That, the brains behind Clearview AI, whose facial recognition app hit the headlines in January 2020.

Recall the app's basic recipe. First, scrape billions of images from Facebook, YouTube, Venmo – in fact anywhere we are expected to show our face – and process them with high-level AI to distinguish those faces from each other in ever more precise ways. Then use the resulting metadata to support someone else's decision process. For example, the decision processes of an entity who wants to know whether this face, seen partially from an odd angle, is the same face that identified itself on that Facebook post. Finally, license the recipe to law enforcement agencies in the US and elsewhere, and you have a tool that the law could only dream of. There's no evidence incidentally that Facebook, YouTube or any other platform anticipated this use of the billions of smiling faces they host, but then there was never anything in their platforms to stop it.

Ton-That saw nothing wrong at all with what he did. As an enterprising individual who, as he put it, 'grew up on the internet', Ton-That made exactly the same move that, on a smaller scale, Mark Zuckerberg performed as a Harvard student when he downloaded and used without permission his female peers' portraits to create

Facebook's precursor: FaceMash.[102] The lineage is clear, since neither saw this as raising moral issues. As Ton-That put it, this was just 'the best use of the technology'.[103] And yet, Ton-That (and Zuckerberg) were also making the classic colonial move: grab everything because it is just there, ready for the taking.

Let's put to one side the evidence uncovered by the *Huffington Post* of overlaps, to put it politely, between Clearview AI's staff – indeed Ton-That's own social networks – and far-right activists in the US.[104] And let's put to one side, for a moment at least, the shocking implications of Ton-That's techniques for human freedom. Focus instead on a more basic point, that Ton-That's techniques are no more than an effective application of the core techniques of data colonialism: first, capture a data territory, then refine it and generate new data fruits that you completely control, and finally sell access without concern for how this data might be used. Or in other words: Explore, Expand, Exploit and Exterminate.

In fact, one of Palantir's patents describes something rather similar: a face photo on a smartphone is sent to a server that recognises it by reference to a huge database and then automatically links it to all the social media data associated with that face and its owner. Because of Palantir's secrecy, we just can't know if it does anything like that, but we do know that Peter Thiel financed Clearview's early operations to the tune of US $200,000, later converted into equity.[105]

There is absolutely no evidence, however, of Palantir's further involvement with Ton-That's enterprise. Why would there need to be? The idea had its own momentum.

Clearview offered 30-day free trials of its tool to police forces. Police quickly found that Clearview's facial recognition tool drew on a much larger database and had a much more flexible algorithm than those supplied by, say, the state of Florida. After all, Ton-That had leveraged the vast data territory of Facebook and YouTube, to name a few. In any case, Clearview never had to know its tool was accurate, only that it produced results that were credible bases for law enforcement decisions. Clearview's own best claim in 2020 was that it found matches at least 75 per cent of the time.[106] No matter: that was enough to launch Clearview's business to global prominence.

Clearview's unauthorised plundering (we might call it 'pirating') of some of data colonialism's booty generated a panic in some quarters. Legal actions against its data sales to the private sector were taken (with fines imposed by a Greek regulator and the UK Information Commissioner's Office), followed by an outright ban in the US state of Illinois. But that was negative reaction to its 2020 business. When we checked in again on Clearview over two years later, it had rebranded itself with an impressive website, full of reassuring headlines: 'Advancing Public Safety', 'Securing People Facilities and Commerce'. It claimed to be ranked number one for identifying faces in the 'US and the Western World', with an expanded database (20 billion+ faces) and improved accuracy (99 per cent+) 'for all demographics'. It boasted Ukraine's armed forces among its satisfied clients and a slew of positive media coverage. Ton-That's redemption from colonial raider to security provider appeared complete.[107]

But the transition from lone adventurer to a key player in military operations is no more accidental in the age of Big Data than it was when the Americas or India were captured. Yes, the geopolitical context is today very different. We are in the middle of an emerging new Cold War between the West and China and Russia. But what matters most is territorial expansion.

Sometimes, data colonialism's lone adventurers seem benign. Take Alex 'Sandy' Pentland, an MIT Professor who in 2011 was named one of the world's seven most powerful data scientists by Tim O'Reilly, himself the creator of the term 'Web 2.0'.

Pentland does not just do data science. He has developed a whole new version of *social science* to make sense of the sorts of data analysis he does. In his 2015 book *Social Physics*, Pentland proposes to replace existing social science models with the search for 'statistical regularities within human movement and communication' that can generate 'a computational theory of behaviour'.[108] The source is personal data, the 'digital breadcrumbs' each of us leave as we pass through the world. The result is basically network analysis: finding patterns in our social interactions based in the structure of our social networks and the people we interact with.

The goal of all this data crunching? To develop a new sort of social knowledge, built from a macro-understanding of how social networks operate and replicate. Pentland has applied this approach to work organisations: he developed a so-called 'sociometric badge' for measuring the 'culture' of organisations through their networks of human interactions, an idea taken up by McDonald's among others. But he has also applied it to social interactions more generally, and even to whole countries (his 'Data for Development' program collected mobile and demographic data for the whole of the Ivory Coast).[109] Among Pentland's companies are Endor.com, a blockchain-based platform that claims to make AI predictions widely accessible (the sorts of predictive modelling that many businesses now want),[110] and Cogito, a leader in the AI-driven inferences that can be made from people's voices.[111] Although in recent work Pentland has focussed on how data capital can be placed at the service of communities,[112] it remains unclear whether Pentland is willing to question the imperative to extract data from everyone and everything that lies at the heart of the new data science.

Whatever its arguable benefits, Pentland's entrepreneurship illustrates even more clearly than Ton-That's the emerging shape of a new science for the data colonial era. Just as botany and zoology once emerged in the course of colonising nations learning how to rule their new lands, so today we have 'social physics', the science that claims to map the phenomena of the data colony's vast archipelago.

We the Consumers

The geography of those responsible for colonialism has always been complicated. After all, the home populations of colonising nations such as Britain had a large stake in colonialism, even though they may have felt themselves far from the realities of what empire did in its 'distant' colonies. Colonialism fuelled massive economic growth back home, as recent campaigns for more awareness of Britain's massive and long-continuing economic reliance on the slave trade have reminded us.[113] Today, there is still exploitation at a distance linked to data colonialism: workers from the Global South who do the underpaid manual labour of preparing materials to train AI models, or the labour of extracting the minerals needed for our electronic devices. But the geography appears very different from that of historical colonialism, because its general target is everywhere: human life (assuming it enters, willingly or not, into data relations).

How data colonialism affects people depends on where they are, their social status, what sort of work they do, and how each is linked to the longer histories of power, especially colonialism. Business elites and other elites the world over may be irked to hear that their photo was in a Clearview database and that their platform and browsing data are likely to be profiled, but the net result of data extraction may just confirm their position among the elite. If, however, you are one of the many economically disadvantaged in the Global South or Global North, you will already know first-hand the force of continuous data extraction and how it is used to judge you. You confront this force when you try to get a loan, or access welfare, or when your only possibilities of work depend on you submitting to the surveillance that comes with a job in the gig economy.

Beneath this complexity, there is one clear truth. That all of us, wherever we live, as users of the platforms and apps that increasingly make up daily life, help sustain their regimes of data extraction.

Our continued acquiescence to their terms of service is a key part of data colonialism's deal.

The question is often posed to users in purely individualistic terms: *my* data is or is not at risk, the portability of *my* photos between platforms needs to be protected. This makes data colonialism seem like a trade-off that each of us can weigh up on an individual basis, for ourselves. But that's doubly misleading. First, as we noted in Chapter Three, the benefits for business come not from us individually giving up our data, but from the fact that we all do, and from the relations between data that flow from this.[114] Second, the transformation of power relations that is under way stems from the fact that all of us are changing our habits in parallel, so data colonialism cannot just be an individual issue. Our social relations with each other are increasingly being shaped by how data is extracted from the lives we live together, as part of the emerging social order of data colonialism.

That said, our participation in this new social order doesn't make all of us members of data's colonial class. That status is reserved for those who exclusively benefit from the exploitation of data territories, and membership of that class is skewed by the inequalities inherited from historical colonialism. Not all old elites have managed to secure their place in data's ruling class (there's no evidence yet that, for example, landed elites in the US or UK are more likely to be data scientists or indeed investors in Big Tech), but they certainly helped to create an unequitable system that the new colonial classes are now exploiting, through new means.

The rest of us – the consumers of digital platforms and the workers in a datafied world – must continuously confirm our acquiescence in the order of data colonialism. We are required to be connected: downloading apps, joining platforms, keeping them updated. Data elites need us and our devices to be constantly plugged into systems of data extraction.

A recent episode rehearsed in a neat political format the basic steps by which, across the world, people are enlisted to participate in data relations every day.[115] In August 2022, in the build-up to India's Independence Day, supporters of the ruling Bharatiya Janata Party (BJP) were encouraged by no less than its leader, Prime Minister Narendra Modi, to put up a national flag outside their house, and post a picture on a special-purpose website. The campaign was called Har Ghar Tiranga (tricolour on every house). Digital rights campaigners were concerned the intentions were not exclusively celebratory. For every picture was required to have a geotag, which in effect tied the location of the supporter's house with their phone number and any other identifying data needed to upload the photo.

Sixty million Indians did this, and received an appreciative certificate from the Ministry of Culture in return. Mamta, a 54-year-old teacher from Uttar Pradesh, was one of them: 'I feel proud, as if I have won a war.' When asked, she admitted she had not thought much about the privacy aspects.[116]

What ideas, what imaginative resources, do we need to develop to resist the enticements offered by data's new colonial class?

CHAPTER 5

VOICES OF DEFIANCE

THE WORST THING for Chunfeng was when customers left bad reviews. 'The delivery guy took too long and my food was cold when it arrived. Plus he was rude when I complained.' Did these people even know what traffic was like at that time? As a *zhongbao* – a gig contractor delivering food in a busy city in China – Chunfeng could have up to half of his day's wages deducted for just one such infraction. That didn't seem fair to him.

Things were better back in the days when the app was just a small start-up, run by a couple of nerdy guys with university degrees. He had quit his job at the factory, tempted by the opportunity, he was told, to set his own hours and be his own boss. Even back then, the money had never been as good as was promised, but he would tell himself he had independence, and he enjoyed driving around the city. Then the company was acquired by one of the largest food delivery apps in China, and things went downhill immediately. Platform fees were always rising, with more and more of the money going to the company (including a portion of Chunfeng's occasional tips). Didn't the owners or the government realise food and gas prices were going up, not down? How was he supposed to support his family?

And yet, he couldn't bring himself to quit. He had worked very hard for that 4.5-star rating. If he left to work for the competition (where things were not necessarily better), he would have to start from zero. So it all seemed rather hopeless, not just for him, but for everyone. A group of drivers had started to meet in person to discuss what to do (they avoided as much as possible chatting about it on WeChat, because they believed algorithms monitored conversations for this kind of talk). What angered them most was to be told that this couldn't be helped, because it was 'The Algorithm' setting the prices. Well, who tells the algorithm what to do? Surely, as he pointed out at a meeting, the algorithm can be adjusted to benefit the bosses or to benefit the workers. There was a lot of anger at the meetings. Things were getting ugly. A driver had set himself on fire in another city to protest unpaid wages.[1] In Indonesia, they heard about some delivery drivers who literally tried to sew their lips shut, in protest against low wages and poor work conditions. Yes, there had been a couple of attempts to protest in his city, but it was hard to get drivers to show up, when it meant losing even more wages. Plus, the police were quite good at dispersing small groups quickly with arrests or beatings.

Then Chunfeng had an idea: let's use the app against them! On the day of the next protest, at the appointed time, everyone would order something cheap through the app and have it delivered to the main square, where the protest would take place. Hundreds of drivers would converge at the same time. Let's see what the police and the algorithm will do then, thought Chunfeng.

Chunfeng's story reminds us that resistance is always possible, even under the most oppressive circumstances. There are always possibilities, even if temporary ones, to turn tools of power, such as a food delivery app platform, against their owners. In this and the next chapter, we consider how to build sustained resistance against something as vast as a new social order built on data extraction. Practical resistance is hardly possible without some sense of *why* one is resisting and that, in turn, depends on something that at first sight might seem distant from practical action: ideas about different futures that challenge the naturalness and inevitability of the path the world is on.

The good news is that we don't need to invent wholly new ideas in order to find inspiration. We can learn from others who confronted analogous problems in the past, especially those historical voices who tried, and sometimes failed, to be heard as they spoke up about the colonial capture of the world's resources, or about the dangers of unlimited computing power. If we want to imagine a world where data colonialism can be effectively challenged, we must rediscover these voices.

Colonialism's Witnesses

Not everyone on Spain's plunder missions was at ease with what they saw. A few were willing to challenge the new colonial system that was meant to benefit them. The best known was Dominican priest Bartolomé de las Casas, who in 1516 was the first to be appointed to the post of 'Protector of the Indians'.[2] He was one of the early voices who spoke against the devastation he witnessed.

Las Casas, born in Seville, was initially part of the Spanish conquest. He arrived in the Caribbean in 1502, and soon became a landowner and slave owner, although from early on he trained as a priest. In time, the pain and injustice he saw, including as a chaplain participating in various colonising campaigns, drove las Casas to a new understanding. He came to believe that a genocide was in the making, and so began a personal war against the emerging colonial system. We know this from his remarkable book, *A Short Account of the Destruction of the Indies*, published much later in 1542 and dedicated to the Castilian king Philip II.

Las Casas's struggle was a frustrating and protracted one, with few victories along the way. Working with like-minded individuals, las Casas petitioned none other than Charles V, the Holy Roman Emperor (Philip II's father), to dissolve the *encomienda* system, in which the Crown issued land titles to colonisers and permitted them to enslave the native population. Las Casas also advocated for laws that would protect the rights and lives of indigenous people. The emperor was somewhat convinced, going as far as passing a set of New Laws in 1542 to govern the New World, suspending all new colonisation campaigns in 1550, and even considering the surrender of control over Peru (though he didn't follow through on that in the end).

Las Casas – as activist, archivist, advocate and public intellectual – was instrumental for a while in shifting public perceptions of the conquest. His greatest opportunity to do so came in 1550, when he was called to Valladolid, Spain, to debate once and for all the question of the natives' humanity. His opponent was the priest and scholar Juan Ginés de Sepúlveda, defending the Crown and the Church. Against Las Casas's arguments for the humanity of the 'indians', Sepúlveda's main point was that colonisation was a kind of just war, sanctioned by God, country and man because the 'indians' were barbarians, not humans.

In the end, the committee appointed by the Church to issue a verdict on the humanity of the colonised refused to do so. Political and material interests won out and the colonial expansion continued. The colonisers learned how to co-opt the language of critics like las Casas to make extractivism more palatable. New laws to govern the colonies, issued in 1573, shifted their language from that of 'conquest' to 'pacification'.

And yet las Casas's voice lives on. We can still read his eyewitness account of what he saw, and still hear not only his fury, but also his clear-sightedness about the moral evils of colonisation.

But it is to colonised peoples themselves that we must turn for the clearest vision. As anthropologists David Graeber and David Wengrow point out in their remarkable book *The Dawn of Everything*, there were, in fact, from the beginnings of historical colonialism, strong and articulate voices among colonised peoples who challenged the colonisers and their claims to legitimacy. It is only recently that Western scholarship has started to take those voices seriously. Until now, those voices tended to be heard in the form of reports by European eyewitnesses or versions penned by European theorists, and for too long contemporary scholars did not believe in their authenticity.[3] As a result, Graeber and Wengrow argue, not only has history ignored the actual impact indigenous critiques had on European colonial societies, but philosophy has never acknowledged the role that indigenous voices played in formulating what became known as 'Enlightenment' values. Let's not repeat that mistake in relation to the choices contemporary societies must make about data.

A remarkable example of that tradition of indigenous critique is the musician, writer and academic, Leanne Betasamosake Simpson. She is a Michi Saagiig Nishnaabeg and member of the Alderville First Nation in Canada, and the author of many books and a recent music album, *Theory of Ice*. There is no way we can do justice here to the breadth and depth of her critique not just of colonialism, but of the damage that capitalism and Western ways of living are doing today. But we want to signal, drawing on her recent book, *As We Have Always Done*, some of the challenges she poses to all forms of colonialism.

One thing on which Simpson is very clear is the landgrab at colonialism's heart. She discusses how, during a university-sponsored research project on the land use of indigenous people of the Great Lakes region, she met the Elders of that area to build a map of their communities' use of the land. The very process of map-making and recording the key events in those communities' history brought to the surface the sheer extent of the losses they had endured. As she writes:

Standing at the foot of a map of loss is clarity. Colonialism or settler colonialism or dispossession or displacement or

capitalism didn't seem complicated anymore . . . It seemed
simple. Colonizers wanted the land. Everything else, whether
it is legal or policy or economic or social . . . was part of the
machinery that was designed to create a perfect crime – a
crime where the victims are unable to see or name the crime
as a crime.[4]

But for Simpson it is not just the historical seizure of land that matters, but the way of thinking that went with it:

> Colonialism and capitalism are based on extracting and
> assimilating . . . The act of extraction removes all of the
> relationships that give whatever is being extracted meaning.
> Extracting is taking. Actually, extracting is stealing – it is
> taking without consent, without thought, care or even
> knowledge of the impacts that extraction has on the other
> living things in that environment.[5]

This very act of seizing territory, without consent and without care
for the consequences, indicates a dysfunctional relation to the world
and a way of living with people that is extractive rather than recip-
rocal. We have found similarly asymmetrical ways of dealing with
people in data colonialism, too, even though the details are inevitably different.

Simpson has translated her ideas about colonialism into practice. In 2012 she participated in the 'IdleNoMore' campaign, which
used digital tools to challenge the continuation of settler colonialism in Canada.[6] But she also, from this experience of practical
action, reflected on the limits of online-only campaigns, and on the
fact that movements not built on face-to-face communication risk
creating more value for digital platforms than for the people
involved.[7]

In a remarkable passage, Simpson articulates the practice of communal knowledge-making and its ethics:

> Meaning, then is derived not through content or data or even theory in a Western context, which by nature is decontextualized knowledge, but through a compassionate web of interdependent relationships that are different and valuable because of difference.

On this social view of knowledge, each person must not only develop their own ideas, but share those ideas with others in a way that tries not to 'interfere with other beings' life pathways', that is, in a way that does not impose her own understanding on others.[8] In this indigenous way of understanding ideas as what emerge in the context of living together on shared land and territory, a completely different understanding of *shared knowledge* emerges. Within this perspective, data colonialism's driving principle – that you can get knowledge by extracting patterns at a distance from myriad data sources across the planet, many of them accessed without consent and outside of rich human relationships – becomes both nonsensical and alarming.

As an indigenous person, Simpson thinks differently about resistance, offering a serious practical vision of what resistance practically involves. Data colonialism may still be in development and so, as yet, fall short of the terrible accumulated violence of five centuries of settler colonialism, which is Simpson's main target. There is no denying this. But while remaining aware of what Simpson's ideas mean for her specific struggle, we can still apply them to the different but related context of data colonialism. For Simpson, it is a matter of rejecting as a whole the civilising narratives that colonialism offers, rethinking and reworking how we live, not merely adjusting this or that isolated practice.[9] At the core for Simpson (and also perhaps las Casas) resistance means holding on to values, even in the most difficult circumstances: values that, as Simpson and many other indigenous thinkers have advocated, emerge not from abstract thought of lone individuals, but from shared life in a particular territory.

This is an important thread that will run through this chapter. For it is such socially generated values with which we risk losing touch when our lives become ever more layered with data territories, imposed by distant platforms and AI systems.

No Modernity without Colonialism

We are in the middle today of a revolution in how we understand the world, a revolution of which this book about data is just a small part. That revolution consists in grasping, as our new starting point, that all the knowledge developed about the world in the past five centuries has been shaped by the colonial landgrab and its consequences.

To say the least, this is a challenge to mainstream ways of thinking about technology, AI and indeed everything. It is not that the basic facts of colonialism have ever disappeared from view. On the contrary, in colonial nations such as the UK, they have been endlessly celebrated as the triumph of those nations' positive mission in the world!

Today's much-needed revolution of thought, however, means acknowledging two things. First, that what has been imagined as the 'internal' history of European and North American societies was always shaped by a larger global history of violent exploitation of colonial territories. This process was assumed by the home colonising nations to be 'external' to their history, something too far away to matter for their own identity. And second, Europe and North America's apparently privileged role in the past five centuries of world history, including supposedly universal histories of Western culture and science, was the *outcome* of that colonial exploitation, not its justification.[10] Our Western view of history resulted from the West's power to decide how history gets told, a power that flowed directly from colonial dominance and ensured that other histories of what colonialism meant were almost entirely forgotten. This historical erasure is now becoming unsustainable in the Global North, but in the Global South, those alternative stories have never been entirely out of circulation.

Under conditions of historical erasure (themselves the direct result of the power inequalities of colonialism), ideas and indeed remembering history differently take on a directly practical force. Three writers from the diverse areas of economic history, social theory, and the history of ideas, can help us see this more clearly.

A Rebel Historian

Eric Williams was born in the former British colony of Trinidad and Tobago in 1911. He became leader of its majority political party in 1956, the country's first premier in 1962 and its first president when it became a republic in 1976. His preeminent political role ensured him the status of 'father of the nation'. He was therefore, by any measure, a successful fighter against colonialism. But at a time when the struggles of former colonies to claim independence are largely over, it is Williams' ideas that have the greatest practical force.

After completing his doctorate at Oxford in the 1930s, Williams went on to publish a book called *Capitalism and Slavery* that upended the mainstream histories of Britain's Industrial Revolution as narratives of heroic homegrown innovation, with scarcely a mention of the colossal colonial wealth that fuelled it.

Based on extensive documentation, Williams showed that those British mainstream histories were a self-serving myth that bore little relation to the realities of the eighteenth-century English economy. They conveniently forgot the huge benefits England derived from the notorious triangular trade that linked (1) the selling of English manufactured goods to Africa to (2) the supply of slaves to English colonies in the Americas and (3) the subsequent export of staple crops (cotton, sugar and tobacco) back to England. Williams quotes an English economic commentator of the time who was not embarrassed to say that the British Empire was 'a magnificent superstructure of American commerce and naval power on an African foundation.'[11]

Williams went further and challenged the idea of colonisers that African populations were enslaved because of their unique physical suitability (supposedly, only they could physically endure the gruelling work of the plantations). The explanation for who became slaves was, according to Williams, much simpler: the economics of cheap labour. Yes, there was an important precondition, that the indigenous Caribbean population had been mostly decimated by the early effects of colonial arrival. But once slaves were imported to the Caribbean from Africa as replacement labour, they were 'just there' for plantation owners to exploit, a pattern that would be replicated in the colony that later became the United States. According to Williams, then, Africans were not racially predisposed to slavery; they just happened to be the cheapest available source of mass labour needed to exploit cash crops.[12]

In these various ways, Williams was anticipating current debates about the relations between capitalism and colonialism by decades. He punctured the fictional idea that the English Industrial Revolution resulted from the miraculous innovation of white English entrepreneurs. Brute force and basic economics were, as he showed, much more important, particularly the economics of colonialism. While Williams' statistics on the economic contribution of slave labour to eighteenth-century England have been debated, his challenge both to economic history and the history of colonialism/capitalism has been crucial. His message: there can be no capitalism without colonialism, or without the subjection of labour, including via slavery.

The myth of capitalism as purely an 'internal' Western development has taken even longer to dismantle, since it also meant taking
seriously the implications of the US's nineteenth-century slave
plantations. Already in the 1930s, the Black sociologist W.E.B. Du
Bois had analysed very clearly their corrosive consequences for contemporary American society, but his work was almost entirely ignored
by mainstream history and sociology. It has taken until the last two
decades for a movement of historians to excavate the central role
that colonialism and slavery played in the growth of the US and UK
economies in the nineteenth century – indeed in the emergence
of a global economic system. The result is a new understanding of
capitalism as never separate from colonialism, a capitalism, in the
words of the historian of cotton Sven Beckert, based on, 'slavery,
the expropriation of indigenous peoples, imperial expansion, armed
trade, and the assertation of sovereignty over people and land by
entrepreneurs.'[13]

Why has it been so difficult for Western societies to see capitalism's ongoing involvement with, and dependence on, colonialism?
Eric Williams ended his country's colonial status back in 1976 and
yet, today, a quarter-way through the twenty-first century, we are
still struggling to grasp these points. What if this blindness is itself
an ongoing cultural effect of colonialism? If so, there must be something more enduring to colonialism than just its political structures.
But what exactly?

The Coloniality beyond Colonialism

For an answer to this question, we must turn to another voice, the
Peruvian sociologist Aníbal Quijano, and his concept of coloniality.
Writing more than thirty years ago, Quijano wanted to understand
what linked the practical side of colonialism (the conquest of territory, the pattern of low-status work in colonial 'peripheries' and
higher-status work in colonial 'centres') with the ways of thinking

that accompanied colonialism, above all its racism. Remember the idea that Sepúlveda presented, in response to las Casas, that the colonised peoples of the Americas were barbarians inferior to Europeans, a lower form of humanity that deserved nothing better than to be conquered and ruled.

In contrast to Eric Williams' purely economic explanation of why slavery was targeted at peoples of African descent, Quijano argued that, from early on, the everyday functions of colonialism depended on a distinctive way of imagining humanity and its capacities for knowledge. The way power was exercised under colonialism by particular races *over* others was itself for Quijano not an aberration, but something distinctive about the modern way of understanding the world. It was this corrosive idea that made colonialism more than just an unending sequence of senseless violence. With this new 'civilising' story of inferior races that deserved to be colonised, colonialism as a project gained momentum and legitimacy. Quijano called this understanding the '*coloniality* of power'.

The coloniality of power captures two key ideas embraced by colonisers. First, that there is one and only one way of rationally organising the world, which requires the distinctive way of producing knowledge associated with what we know as 'Western science'. Second, that this form of knowledge production authorises a hierarchy of races (even if it has become increasingly difficult to say this explicitly). This hierarchy differentiates between those who are closer to rationality (white men, supposedly) and those who are farther from it (non-white people, we are told). The 'inferior' races (for example, indigenous groups) become merely the objects of the knowledge produced by the Men of Science, who overwhelmingly come from Europe and North America.

Obviously, it is Quijano's whole point to insist that this is a debased argument. And yet, if we are honest, it captures not only many of the subtle dynamics around race in contemporary life, but also why many of the injustices in algorithmic decision-making discussed in Chapter One get treated somehow as natural and unavoidable. Indeed it gets to the core of why the language of Big Tech and AI today seems so powerful, even when its flaws are becoming ever more obvious.

Today's practices of Big Data and technology-driven decision-making seem on their surface to have nothing to do with race, gender and class. But what if the era of Big Data, with its new asymmetrical power relations, is generating new ways of thinking hierarchically about human beings?

A key implication is suggested by some reflections Quijano made about how to resist coloniality. Not, he says, by rejecting knowledge and rationality, as if they were tainted for ever through their exclusive association with the West. The best way forward for resisting the coloniality of power is to develop a better, non-hierarchical idea of knowledge and rationality, for there is no way of relating to the world that does not depend on *some* idea of knowledge, *some* idea of rationality.

What would this look like? Quijano proposes an inclusive vision of knowledge which aims to encompass, rather than to exclude, important differences in how different groups of people view the world. In fact, Quijano insists that in most cultures other than the West, knowledge is already understood as built from diversity of perspectives, rather than by denying diversity. On that view of culture, it is the inclusiveness of a way of understanding the world that qualifies it as real knowledge, and not just a personal perspective on the world.[14]

Two things are striking here. First, this apparently abstract account of knowledge has a lot in common with indigenous visions of how knowledge is produced in communities (remember Leanne Simpson). Indeed, other theorists of coloniality such as the anthropologist Ramón Grosfoguel have criticised Quijano for not explicitly recognising the voices of indigenous thinking that are audible in his work.[15]

Second, this more inclusive view of how real knowledge gets produced has, again, huge practical implications in a world like ours where the practices of Big Tech and Big Data offer a directly opposed top-down institutionally driven view of knowledge. This is not just a matter of people lacking the technical expertise to understand the impenetrable language in which most explanations of algorithmic workings are generally couched. Imagine you wanted to challenge how Facebook's algorithm works, or how Amazon or TikTok categorise your desires, or how a social welfare system's algorithms operate. You will have great difficulty getting answers, because, even if the algorithms were disclosed to you and you understood the details, your questioning of their rationality would be regarded from the start as naïve and uninformed. Big Data's knowledge is the knowledge of institutions, not people, and in this it perpetuates exactly the feature of Western science that makes it, at a deep level, colonial.

If this is the case, rejecting data colonialism may mean challenging the view of human knowledge that has been at the heart of colonialism for five centuries, by insisting on the continuing connections between real knowledge and live communities of practice. This idea will be of major importance when we turn directly to practical resistance in Chapter Six.

Reimagining Humanity beyond Colonialism

Another writer who helps us think differently about knowledge, science and rationality is Sylvia Wynter. Wynter was born in Cuba, grew up in Jamaica, and has lived much of her life in the US, where she is Professor Emerita of Spanish at Stanford University. Her work has been devoted not just to defining 'coloniality', as Quijano did, but to unsettling and overcoming it. Like Quijano and Williams, her core idea was itself deeply practical: to resist head-on dominant Western views of history shaped by the victors of colonial struggle.

For Wynter, there is no point in abandoning knowledge and rationality, just because their Western forms are tainted by the history of colonialism. If a hierarchical and racialised view of knowledge has become dominant, it is our job to develop a more inclusive view.[16] Taking inspiration from her, it would be similarly absurd to abandon the idea of data; instead, we must rethink it within a more inclusive view of how human beings know the world.

In other words, the point of critical thinking is not just to reject but to replace an unsatisfactory view of human knowledge. In that spirit, we need a process of thinking about data that includes a wider range of people and perspectives. Why continue with data practices that simply reproduce, in one way or another, the inheritance of colonial and racist thinking? This is exactly the movement that critical data scientists have recently initiated, inspired by the writings of information scientist Safiya Noble, sociologist Ruha Benjamin, and others.

Most interesting of all is Wynter's attempt, from the perspective of the history of ideas, to reinterpret what happened in 1492 – the year Christopher Columbus 'found' what became known as the Americas – and the decades of debate at the Spanish court that followed. As Wynter points out, the discovery of a continent that was unknown to Europeans was a huge challenge to the authority of the Catholic Church, coming only two decades before Copernicus's discoveries further undermined earlier understandings of Earth's role in the cosmos, and it was this that generated the debate between las Casa and Sepúlveda that we mentioned earlier. For Wynter, it was las Casas's interlocutor Sepúlveda – confidently insisting on the privileged right of rational Western man to seize the territories of inferior beings – who represented something radically new: the birth of a secular, but also deeply racist, vision of the world. It was those ideas that won out, and it is the continuation of those arguments at the heart of powerful understandings of knowledge and humanity that must still be countered in Big Data discourse today.[17]

Wynter's proposal for rereading history does not mean reverting back to the worldview that was undermined in 1492. It means taking seriously the historic crisis of the closed worldview that had erupted by the early sixteenth century and finding – five centuries on – new and better solutions. That requires, once again, an inclusive, rather than hierarchical, view of humanity.[18] If we follow Wynter, our goal becomes not to abandon thinking about what things like Artificial Intelligence mean for human knowledge, but to look for ways of *reimagining* AI and Big Data which genuinely move beyond the colonial ways of thinking that, until now, have shaped them.

Warnings from an Earlier Computer Age

To take on that challenge, however, a very uncomfortable truth must be confronted: that AI and Big Data are already retooling coloniality (the thought-system of colonialism) for an age of continuous surveillance, instantaneous communication, and limitless management power.

If that is true, in order to challenge today's new forms of colonial thinking it is not enough to reinterpret past history in general terms. It is essential to reinterpret the history of the computer technologies that have made today's AI and Big Data possible. To do this, it helps to take a step back into computing's almost forgotten past. One of the mathematicians who helped invent the computer was Norbert Wiener. Rather than simply celebrating this new technology, Wiener was full of foreboding. In the original 1947 preface to *Cybernetics*, one of the most revered texts in the history of computer science, he wrote:

It has long been clear to me that the modern ultra-rapid computing machine was in principle an ideal central nervous system to an apparatus for automatic control ... Long before Nagasaki and public awareness of the atomic bomb, it had occurred to me that we were here in the presence of another social potentiality of unheard-of importance for good and for evil.[19]

Those words are chilling, especially when we remember that he was writing just two years after the dropping of atomic bombs on Japan. Wiener is saying that the development of computers is a comparable danger.

Of course, he could not have anticipated the world of connected computers which has launched today's Big Data, digital platforms and Artificial Intelligence (the 'rapid' computer he wrote about was a mainframe that occupied a whole room and issued an answer in days rather than micro-seconds). But even back then, what worried him was the social potential for using computers as tools of control, as the means to rule and dominate others. In other words, his fears had already anticipated the concerns that data colonialism raises today.

And yet, in spite of Wiener's celebrated status as a founder of the computer age, these fears were ignored. No one heeded them when networked computers started to scale to wider society in the late 1980s. No one heeded them when control of the internet passed from public hands to private corporations in the early 1990s. And no one heeded them when social media platforms in the early 2000s were designed as data territories under those platforms' exclusive control.

If Wiener's warning has been remembered, it is only very recently and in a rather rarefied context: the development of very large-scale AI, the sort that underlies ChatGPT and other recent developments. Stuart Russell, a global AI expert and BBC Reith lecturer, recalls Wiener's concerns in his recent criticisms of where

Artificial General Intelligence (AGI) is heading.[20] Wiener had not been afraid to use moral language. He wrote of computers' 'potentially unheard-of importance for good and for evil'. It's exactly Wiener's moral reference point, Russell argues, that we must recover, if we are to correct recent developments in AGI that seem driven only by technological rationales. Large-scale AI is expanding to levels of complexity that we struggle to understand in terms of human values and thought-processes, potentially taking away the capacity to intervene at all.[21] Novelists like Ian McEwan and scientists like the late Stephen Hawking have echoed Russell's fears; it was Hawking who said that 'the development of full artificial intelligence could spell the end of the human race'.[22] Wiener's writings help us see the kernel of truth in the chorus of recent calls by leading AI experts to 'halt' its most advanced forms.

But Wiener's warnings didn't end there. After the passage we cited, he expressed a concern that computer machines could take over human work, a fear that had been expressed by late nineteenth-century novelists such as Samuel Butler. Wiener was agnostic about whether such computerised task-saving was good or bad for humanity. But of one thing he was sure:

> It cannot be good for these new potentialities to be assessed in the terms of the market, of the money they save . . . The answer, of course, is to have a society based on human values other than buying or selling.[23]

On the chances of this, and the chances of avoiding the social evil from computing he feared, Wiener was pessimistic:

> We do not even have the chance of suppressing these new technical developments . . . The best we can do is to see that a large public understands the trend . . . [T]here are those who hope that the good of a better understanding of

man and society which is offered by this new field may anticipate and outweigh the incidental contribution we are making to the concentration of power . . . I write in 1947, and I am compelled to say that it is a very slight hope.[24]

Wiener feared computers becoming institutions of social control, managed by market mechanisms with little regard for human values and directed only to the acquisition of power.

For sure, Wiener's fears were shaped by the then recent history of Nazism and fascism, and perhaps the politics of Soviet Russia too (less than two years later, George Orwell published the novel *1984*). But it is wrong to dismiss Wiener's fears as outdated. Not only has the renewed relevance of Orwell's work to our surveillance-saturated societies been noted recently, but Wiener's fears have echoes in the massively unequal power relations of today's data territories.

Early Apprehensions about AI

A small group of computer scientists in the 1970s and 1980s confirmed Wiener's concerns about the social costs of expanding computing power in an uncontrolled way. Typically for the time, these voices were white and male. That caveat aside, we still have much to learn from them. So, let's turn to one of those voices: Joseph Weizenbaum, an MIT professor and refugee from Nazi persecution who, after work on computer systems for banking, became famous in the 1960s for designing one of the first natural language processing systems.

The programme was called ELIZA. Paradoxically, Weizenbaum's concerns about the direction of computing arose not from the failures but from the apparent successes of ELIZA. The program attracted attention when it proved capable of interacting with a real person in a therapeutic situation, not only giving the impression of understanding them, but also offering some advice. ELIZA, it seemed, had triumphantly passed the famous Turing Test, which asked if a computer-generated output would lead a human to think they were communicating with another human. When the transcript of the therapeutic 'conversation' was published, the 'advising' computer quickly gained the nickname DOCTOR. But Weizenbaum was horrified: far from seeing the script as just a basic demo, scientists read it as a blueprint for automated psychotherapy (think of the time and money society could save!). If scientists couldn't understand the limits of computing in such an obvious case, Weizenbaum feared there was a problem with the 'scientific outlook' itself, a problem he set out to investigate.[25]

For Weizenbaum, it was obvious that there were limits to what computers could do. Computers, after all, are machines which apply rules: rules that only work if they are absolutely *explicit*. Human beings, by contrast, understand things by relying on context, much of which remains *implicit*. There must therefore be acts of understanding that humans (indeed, non-human animals) can perform based on contextual understanding, but computers can't. We need to understand this, before relying on computers too heavily.

Weizenbaum was also interested in a different kind of limit: moral limits. As a computer scientist, he understood that a computer's dependence on explicit instructions necessarily imposed limits on how it would function in a social setting:

> [A] computing system . . . accepts only certain kinds of 'data'. Which means that relying on such a computing system for social tasks means clos[ing] many doors that were open before it was installed.[26]

Yet some of these doors that computing closed off might be humanly important. If so, we have to ask not just what a computer can do, but what it ought (or ought not) to do. As Weizenbaum put it, ' "can" does not imply "ought" '.[27]

What increasingly worried Weizenbaum (and led him to write the book *Computer Power and Human Reason*) was that so many scientists ignored these inevitable limitations of computers. Mainstream computer science, according to Weizenbaum, was too interested in control: not the broad social control Wiener feared, but the narrow instrumental control that operating a computer gave programmers. Yet this narrower type of control also had social consequences: it was a form of power that Weizenbaum believed to be spreading through society. The final chapter of his book was called 'Against the Imperialism of Instrumental Reason'. A new sort of social order was being built, he feared, that required us to place moral limits on our use of computers.

Perhaps, Weizenbaum wondered, 'there are some acts of thought that *ought* to be attempted only by humans'.[28] And, though he did not talk of racism or colonialism exactly (yet he does mention the word 'imperialism'), the reason for Weizenbaum's concern had something in common with the questions Quijano and Wynter were raising about the coloniality of power: 'these [computer] systems . . . have reduced reason itself to only its role in the domination of things, man and finally, nature.'[29] Weizenbaum's fear for the type of society that will result if computer power is used in this narrow way takes us back again to Simpson's idea about the dangers of decontextualised knowledge. If knowledge emerges outside of a human context, we risk an irresponsible type of knowledge, exactly the fears now being voiced about AI.

Weizenbaum expressed those fears for a world of computer connection unimaginably less developed and commercially driven than ours. And his core idea retains practical force today: the idea that computers are, in the end, just tools that can and should be brought back under human control, if managed in accordance with human values.[30] It was only the narrowly instrumental way of thinking about computers that made us forget this.

What do these voices from the early history of computing tell us? That the data colonialism thesis, far from being outlandish, echoes exactly the worries about the unrestricted uses of computer power that leading figures in computer science have been voicing from the beginnings of computer science.

Imagining the Battle to Come

Realising this common theme in the history of computing and its common ground with much broader debates about colonialism takes us one step closer to imagining what resisting data colonialism might mean in practice.

The work of historian and theorist Achille Mbembe is helpful in this regard (he was born in Cameroon, and now lives in South Africa). In earlier work, Mbembe provided a clear diagnosis of the continuing effects of colonial history in what he called the 'post-colony' of Africa, countries where opportunities, resources and political structures remain profoundly marked by centuries of colonial extraction (by Britain, France and others). Mbembe unpacked the many levels of violence that colonialism involved: the initial seizure of the land, the violence that stabilised colonial power, and the violence that sustained colonial power over the long term. This violence was both physical and symbolic, and it still leaves a profound mark on the organisation of contemporary African societies.[31]

In his more recent work, Mbembe has turned to thinking about the future: the future of Africa and of the world, a world profoundly shaped by technology. He is the opposite of naïve when it comes to the continuing power and influence of the so-called Global North across the world today. But our starting point for thinking about the future (future resistance, future solutions) cannot, Mbembe insists, be Europe or the Global North. Europe, he writes, 'is no longer the center of gravity of the world. This is the most significant event, the fundamental experience, of our era.'[32] Since the violence of colonialism is continuing, the challenge and resistance must be framed from perspectives outside the Global North. But how?

When Mbembe says that the violence of colonialism is continuing, he does not just mean that the neocolonial legacy continues in the sites of historic colonialism – of course it does. Mbembe is particularly concerned by new forces of data extraction driven, not least, by Silicon Valley, and targeting human life everywhere. The transformation implied by the endless extraction of information is, in Mbembe's view, affecting the Global North too, or at least those outside their global elites. He is worried by broadly the same developments that we have been discussing in this book: new territorial relations of data extraction, extended management power, increasing surveillance, the unthinking expansion of AI. Mbembe calls this 'the becoming-artificial of humanity'.[33]

These new developments, he argues, have a profound implication for how any of us anywhere can imagine resistance. His way of putting this is provocative:[34]

> [There is] the very distinct possibility that human beings will be transformed into animate things made up of coded digital data. Across early capitalism, the term 'Black' referred only to the condition imposed on peoples of African origin . . . Now, for the first time in human history, the term 'Black' has been generalized. This new norm of existence . . . expanded to the entire planet is what I call the *Becoming Black of the world.*[35]

What does Mbembe mean by this phrase? He is not referring exclusively to skin colour, and he is not only referring to racial identity. When he says the term 'Black' has been 'generalized', he is referring to a condition of unfreedom imposed on Black people for centuries, but now extended to 'the entire planet' through the datafication of human life. Seen from the decolonial framework we have advanced in this book, we can reinterpret the development to which Mbembe refers as the condition of data colonialism itself. It is through this that, as Mbembe says, 'capitalism sets about recolonizing its own center'. In this transformation, no region is spared. Resistance everywhere becomes a matter of working out what 'reserves of life . . . [can] escape sacrifice,'[36] not only in the Global South, but potentially everywhere.

But how to resist colonialism if we understand it in this way? We need to focus around a common question. But once again, we can't invoke a 'common humanity' and ignore the continuing forces that until now have divided humanity and reproduced inherited colonial hierarchies. So for Mbembe, like Wynter, the task is to rethink what we mean by humanity. Given the divisive legacy of colonialism, there is only one way to do that, and that is in the context of the still larger common challenge which overshadows his argument: the challenge of inhabiting together a planet under threat of destructive climate change.[37]

It is at this point that we need to turn to Canadian author and activist Naomi Klein, who has written with great force about both climate change and the continuing need to confront the violent legacy of colonialism. We cannot possibly do justice here to the range of Klein's influential writings. But let's focus on one core point, which goes to the heart of what resistance to data colonialism is about.

Klein's work has not usually focussed on data. But an important exception was Klein's intervention early in the pandemic in reaction to New York's Governor Andrew Cuomo's enthusiastic adoption of plans by the ex-CEOs of Microsoft (Bill Gates) and Google (Eric Schmidt) to introduce more tech – and much more data extraction – into the everyday life of New York's citizens.

Klein had made headlines a decade earlier when she spotted the pattern in how big business of all types, and powerful governments, used disasters to create opportunities for the private and public sector to exploit vulnerable populations more ruthlessly. She called this 'the shock doctrine' of 'disaster capitalism'. Its ethos? Always use a temporary crisis as leverage to install permanently the future models of value extraction that benefit you, if you are a corporation or government. The first example was the imposition of the very first neoliberal economic regimes in Chile, after the coup against President Allende in 1973.[38]

Klein saw a similar pattern in New York City in the early months of the pandemic, and the core of this pattern was unlimited data extraction. Called 'visionary' by Governor Mario Cuomo, Eric Schmidt's and Bill Gates' plans were to introduce systems for remote teaching, remote work, remote health provision, and so on – in other words, systems for extracting value through data and data-related services. The supposed goal was to fix the temporary, if extremely serious, problems caused by the Covid-19 lockdowns. As Cuomo put it, 'all these buildings, all these physical classrooms – why, with all the technology you have?' Why, that is, would you even need them, if you had the infrastructure to allow people to work and study remotely?[39] As we know, this shock doctrine brought with it increased extraction and increased surveillance. 'Pivoting' to tele-presence was a great business for the tech sector, one of the few that experienced record profits during the pandemic. Not as fortunate was the working class, the so-called 'essential' workers, for whom going remote was not a possibility.

In this context, Klein is challenging exactly the vision of unconstrained data extraction that we saw was a core vision of data's colonial class. But the way she frames this argument connects with her earlier work on the environmental crisis. In her acclaimed 2014 book, *This Changes Everything*, she notes that climate change will not be addressed until we stop the profit-driven extraction practices, particularly of fossil fuels, whose sales drive the consumption that results in global warming. Political incompetence aside, Klein argues, a key reason for the failure to address this is that, from the perspective of the extractors – and the populations in the Global

North and elsewhere who condone them – the damage is always happening *somewhere else*: in a distant country, or even in places in one's own country that somehow don't count. This *'somewhere else'* is not just where the extraction happens, but also where climate consequences are most acutely felt. Klein calls these ignored places 'sacrifice zones'. Colonialism has been particularly good at creating them.

As Klein writes, 'extractivism became rampant under colonialism because relating to the world as a frontier of conquest – rather than as home – fosters this particular brand of irresponsibility'.[40] This idea of extraction carried out with impunity at a distance was perfected under capitalism, but it was already there at the origin of historical colonialism. Historically, there have always been constraints on where extraction could happen (difficulties of particular mining locations), or how much profit could be generated (for example, variable transportation costs).[41] Without these constraints, extraction would have been even more extreme, for the *idea* of extractivism, especially when carried out at a comfortable distance, knows no limits.

This opens up a frightening possibility for the case of data extraction, which Klein hinted at in her piece on the pandemic 'screen new deal': the sources of human data are *not* far away from the corporations that want them, they are just a device away. Switch your device on, and you are already plugged directly into an external system that can seamlessly extract data from you and your life. The friction of distance, which once constrained extraction during historical colonialism and capitalism, is now reduced almost to nothing. The life sources from which Big Tech wants to extract data – all of us digital consumers – are right there on the other side of a screen. So, Klein warns us, if we let ourselves be misled by those data-extracting narratives, we may give up some things that are pretty important for our whole future as human beings. Not least our autonomy.

And this is exactly what we mustn't do. As things turned out, the adjustments many of us made early on in the pandemic – towards remote working and ever greater reliance on data-extracting platforms – were not converted into completely new ways of organising work. Although slowly and with resistance, we have, most of us, got back to working together face-to-face, at least some of the time.

But suppose the uncertainty of the early pandemic had continued much longer, due to delays in developing vaccines. Suppose the temptation to buy in permanently to the sort of proposal made by Big Tech leaders Schmidt and Gates had taken over, propelled by the belief, as one industry commentator put it, that 'humans are biohazards, machines are not'. The 'shock doctrine' (borrowing Klein's term) of data extractivism might then have locked us into patterns of remote working, teaching and surveillance for the future,[42] committing ourselves to a world, as Klein puts it, 'in which our homes are never again exclusively personal spaces but are also, via high-speed digital connectivity, our schools, our doctor's offices, our gyms, and, if determined by the state, our jails'.[43] In fact, as we've seen in earlier chapters, that is exactly the future that many people do indeed seem to have accepted: a future whose costs will be extremely unevenly distributed, and whose environmental costs will certainly be high. We really must, as Klein insists, start saying no.

Through a series of steps between apparently abstract ideas, we have reached the point where we can start to imagine what resistance to data colonialism could practically look like.

Resources for Resistance?

Ideas are tools. We manage our lives differently when equipped with new ideas. In this chapter we have unlocked some new ideas that can arm us against the worst of data colonialism. Let's sum up what we have learned.

- Never give up on the basic values of what we believe to be acceptable or unacceptable (las Casas).
- Never accept new notions of knowledge whose apparent usefulness depends on overriding the world of human inter-relations that we already trust and value (Simpson, Klein).
- Rethink our understanding of what counts as knowledge and rational behaviour. Reject claims to knowledge that, at root, reinforce inequality and domination. Understand what knowledge is and where it comes from (Quijano, Simpson, Weizenbaum).
- Rethink humanity as a shared project against the background of the accumulated violence of centuries of colonialism (Wynter and Mbembe).
- Acknowledge the special dangers that follow when we use computers as if they had no social consequences, and as if we do not have the right – indeed the duty – to set limits on their usage (Wiener, Weizenbaum, Mbembe).
- And finally: be ready to see colonial extractivism, old and new, for what it is: a grab for power, happening right there in front of us, in the banal details of everyday life (Klein).

With these tools in hand, we are now ready to start thinking in detail about what resistance to data colonialism might look like.

CHAPTER 6

A PLAYBOOK FOR RESISTANCE

JOSEFINA LUCÍA IS a trans-woman programmer, an activist for dissident technology, and a teacher living in Buenos Aires, Argentina. In a country where the life expectancy for a trans person is 35 years, and only 18 per cent of trans people obtain formal employment, Josefina Lucía and her friends are determined to make a difference. Together, they helped start Alternativa Laboral Trans (ALT), a worker cooperative specialising in technology and owned mostly by trans and non-binary people.

ALT's mission is to expand job opportunities for this vulnerable community. To accomplish this, ALT offers website design and development services, and conducts workshops on programming and the application of technologies that are not based on extractivist models. When building a website or app for themselves or for their clients, ALT works with technologies that allow trans people to keep their creations under their control as much as possible – for instance, by using open-source software instead of commercial software. ALT also acts as an incubator for tools such as Lupa (magnifying glass, in Spanish), an app for reporting and visualising institutional gender-based violence which functions as an educational and awareness tool.

There is a tangible need for these kinds of tools, given the high rates of violence that is directed towards gender non-conforming individuals, and the ways in which historically online platforms have sought to force these individuals to conform by limiting the options available to them to define their gender. ALT is also working with the Wikimedia Foundation to develop a platform that will help First Nations people store and share geolocation data about objects and sites that are culturally significant to them.

We have been using fictional composites based on non-fictional details at the start of every chapter, but in this case Josefina Lucía and ALT are one hundred per cent real (see altcooperativa.com). People like them and the others discussed in this chapter are proof that data colonialism can be resisted. For that, we don't need to start with grand, world-changing social movements (although we might eventually get there). We can start with small local acts that demonstrate that, as for Josefina Lucía and ALT, movements to decolonise our data are often tied to other social justice movements. Together, they prove that there are alternatives to the ways extractivist technology corporations, organisations and governments manage the world.

Certainly, data colonialism is not a problem with an easy fix. The countless examples we have discussed throughout the book have made that painfully clear. One cannot simply brush away these news forms of dispossession and inequality with a single new law, a revolutionary technology or even a social revolution any more than we can undo colonialism's five centuries of dispossession and inequality. Data colonialism has unleashed a torrent of social impacts that go well beyond the capturing of data from our daily lives. Let's highlight eight:

1. Workers' rights and their sense of autonomy are being diminished; whole sectors of the economy are being reorganised around data extraction, giving more power to managers and less to workers;

2. algorithms increasingly make important decisions about the lives of the most vulnerable and disadvantaged people in our societies; ethnic minorities, women, gender non-conforming and low-income people have been targeted through new and pernicious forms of surveillance;

3. corporations trying to sell us stuff, and political parties trying to get us to vote for certain candidates or agendas, use data to manipulate us in opaque ways;

4. a global pandemic was used as an excuse to collect more data from everybody;

5. platforms that are more interested in profit than in protecting users' rights have washed their hands of the spread of disinformation and speech that promote hate and violence;

6. data centres, and many other physical aspects of our digital communications environment, continue to deplete the environment of water and energy;

7. the long-term impact of data colonialism on our mental health has barely been studied;

8. all of this, while good uses of data for non-commercial purposes receive little support, with most of the attention and money going to commercial and exploitative uses.

And remember: just because some of us are not feeling the full effects of data colonialism, it doesn't mean that others are as fortunate. Data colonialism, like historical colonialism, is creating an ever more unequal social and economic landscape.

And yet, colonial history has also taught us that it *is* possible to resist. As long as we are willing to make the connection between what is happening with our data and the larger problem of historical colonialism, we may paradoxically make data questions more manageable. We can learn from previous forms of anti-colonial resistance what has worked, and what hasn't. The good news is that, finally, as a way to reject the entire project, an unstoppable force is emerging . . .

Resistance Is Already Here, and Nothing Can Stop It

Resistance to colonialism emerged with colonialism itself. There is already a 500-year history of resistance, even if it is not always visible in our mainstream history books or celebrated in our media.

This history is definitely complex. Yes, colonies became independent nations, but often through violent processes that did not benefit all citizens of those newly formed nations equally. The truth is that anti-colonial liberation movements have been mired in paradoxes, contradictions, and new forms of injustice that replicate some of the old forms of injustice themselves. Nevertheless, out of these complicated histories, inspirational resistance movements have emerged.

These movements have allowed us to assess how colonialism, past and present, is intertwined with poverty, migration, fractured social and gender relations, and even with the impending climate disaster. That is why the wider reckoning with the legacy of historical colonialism, and with the postcolonial growing pains of former colonies, has become a pressing global conversation. While our argument focuses primarily on the field of data extraction, we must recognise how this form of extractivism is connected to others. By rejecting data colonialism head on, we offer one important, but far from exclusive, starting point for addressing the legacy of historical colonialism over the past five centuries.

Confronting extractivism, however, means being honest about one important point: how we are, individually and collectively, implicated in the new social order that is being built through new forms of data extraction. We cannot build resistance without solidarity, but that means being clear from the start about how we have individually contributed to this system in various ways.

In other words, we need to be honest about assigning blame where necessary. The scientists and engineers behind the creation of these extractivist technologies share some of the blame, certainly.

It's no excuse for them to argue that they couldn't foresee the impact that their tools would have, although if that's the case, blame should also be assigned to the professors and schools who educated them, and who neglected to include in their curriculum basic critical thinking tools and notions of ethics and social responsibility. Of course, corporations and states deserve much of the blame, as we have seen throughout the book. But what about shareholders, who demand that those corporations put profits before people? What about the media, which regularly acts as a mouthpiece for the civilising narratives of data colonialism, instead of asking critical questions? And what about us, the users of these technologies? It is one thing to lack knowledge, and another to look the other way when we perceive the impacts of extractivism.

Some clarity about our individual level of responsibility is therefore a good basis for then doing something about the problem. And people are already hard at work resisting data colonialism in many creative and collective ways.

How Citizens Are Resisting

Citizens have exerted pressure on their local governments to curb the use of surveillance systems and AI. In the United States alone, 17 communities have issued bans against the use of facial recognition by police, including Berkeley, Boston, New Orleans and San Francisco.[1] This is a good starting point for resisting data colonialism, since the technology has been shown to be inaccurate (especially for the faces of non-white people) and a threat to privacy, information security, and free expression.

The state of California has gone as far as to ban predictive policing (the practice of using AI and other forms of analytics to identify future criminal activity), and the city of Seattle passed the nation's strongest regulations on surveillance technologies.[2] This continues a trend of citizens around the world taking a stance against some of the most toxic effects of AI and data collection, sometimes adopting more progressive positions than federal governments. One inspiration is the Reclaim Your Face coalition, which helped launch a European Citizens' Initiative to push for more strict regulation of biometric surveillance.[3]

This strategy is not without its challenges, however. According to the most recent data, some of the US cities mentioned above may be undoing such bans in the face of increasing crime rates and pressure from police departments to use 'any available tools' to reduce crime. Lobbying efforts by corporations eager to sell their technological solutions to local governments are also to blame for the ban reversal.[4] Meanwhile, the global movement to build Smart Cities and Smart Welfare systems regularly excludes citizens from decisions about the surveillance technologies that will shape their lives.[5]

Eliminating such invasive technologies will be a long struggle. In the Netherlands, for instance, public pressure forced the government of Prime Minister Mark Rutte to resign in 2021 after a scandal revealed an AI system had falsely accused thousands of parents, primarily immigrants, of committing childcare fraud (Rutte later returned to power, only to resign again in the summer of 2023). Some Dutch cities have continued to use AI to detect childcare fraud, targeting specific neighbourhoods or even individual buildings for surveillance.[6] If data colonialism is part of a social order, as we have repeatedly emphasised, resistance must be for the long-term.

How Workers Are Resisting

What was unthinkable a few years ago is at least a possibility now: there is in the US an Alphabet (Google) Workers Union, an Amazon Labor Union, and an Apple Retail Union, while in the UK Amazon workers are getting support for their unionisation from the long-established GMB Union; and workers at Meta (at least those in the mailroom) are also trying to unionise.[7] The video game industry, notorious for its labour abuses, is seeing a rising union movement.[8] As we noted in Chunfeng's story in the last chapter, workers are also organising themselves against the depredatory practices of gig platforms. From Uber drivers in South Africa to Zomato food delivery contractors in India, gig workers throughout the world are coming together to try to form unions or to exert pressure on governments to pass laws that guarantee their basic rights. Workers in Nigeria are preparing lawsuits against Uber with advice from their peers in the UK, as they all fight a corporation that prefers to deal with workers not as a group, but as disempowered individuals. In the process, transnational worker coalitions like the International Alliance of App-Based Transport Workers (IAATW) are forming.[9] Yes, union membership is still at a record low in places like the US.[10] But these are encouraging signs.

While it is essential to change laws in order to protect gig-labour employees, workers are sometimes taking matters into their own hands and developing DIY approaches to dealing with the platform's algorithms. In India, drivers have formed groups like the Commercial Cab Driver's Awareness and the Telangana Gig and Platform Workers' Union (two of dozens of similar groups that exist on the chat app Telegram) to post questions, troubleshoot problems, and get advice from more experienced drivers. In Indonesia, drivers have used social media to form communities where they provide financial assistance to those experiencing hardship, help each other fix their motorbikes, and even offer 'account therapy' to coax the algorithm to behave in ways that are more beneficial to them. The goal: to try and reverse-engineer the apps to counter their exploitative effects.[11] And in a number of African countries, workers for the online task platform Upwork have found ways to help each other get that crucial first task on the platform.[12] In New York City, a group calling themselves Deliveristas Unidos (made up mostly of immigrants) effectively demanded basic labour rights against the gig economy platforms that exploited them, including a guaranteed minimum wage, transparency about how tips are calculated – one of their slogans was 'Tips are not wages!' – and even the right to use restaurant bathrooms (although ensuring that those demands are met has been difficult).[13] In spite of successes like these, the efforts of United Taxi Workers of San Diego demonstrates how difficult it is for collectives to develop their own technological solutions outside of large-scale capital investment and extractivist infrastructures.[14]

Nonetheless, workers in precarious situations are finding ways of resisting data colonialism. The Homeless Worker Movement in Brazil, for instance, is actively engaged in developing their own approaches to digital sovereignty. For them, this means democratising access to ICT and media production tools in order to 'circumvent unfavorable political structures, mobilize resources, carry out campaigns, promote militancy engagement and document the memory of our popular struggles'.[15]

Meanwhile, corporate and state workers in the Global North are playing a crucial role through their dissent in bringing about change from within the belly of the beast. The cases of government whistle-blowers Edward Snowden and Chelsea Manning, who helped expose the US global surveillance apparatus, are perhaps too unique to generalise. But their courage paved the way for other whistle-blowers and dissenters like Frances Haugen, who revealed Facebook-Meta's willingness to put profit before the safety of its users; or Timnit Gebru, the Google researcher whose pioneering work on the limits and flaws of large-scale AI we discussed in Chapter Three; or Meredith Whittaker, another former Google employee who helped organise a walkout in 2018 that involved more than twenty thousand employees worldwide. Countless other individuals and groups have started to speak out against some of the most powerful corporations in the world, especially their lucrative contracts with government agencies. One of their petitions reads:

We are part of a growing movement, comprised of many across the industry who recognize the grave responsibility that those creating powerful technology have to ensure what they build is used for good, and not for harm.[16]

The courage of these workers, whose lives in various ways are closely entangled with the power of Big Tech, show us that Big Tech is not a monolith and it too can be a site of resistance.

How Activists Are Resisting

The extractivist practices perpetrated by data's new colonial class are no longer invisible or unchallenged. We are witnessing the birth of activist movements against data extractivism that range from local grassroots collectives to well-funded transnational think tanks. NoTechForICE (notechforice.com) was started by the latinx coalition Mijente to highlight the use by the US Immigration and Customs Enforcement agency of the technologies and data services supplied by Palantir, Amazon, Thomson Reuters and others that facilitate inhumane and sometimes illegal methods to persecute immigrants. The NoTechForApartheid movement (notechfora-partheid.com) is a project of MPower Change and Jewish Voices for Peace that works on bringing to light the collaborations between the Israeli government and companies like Google and Amazon to surveil Palestinians (according to Amnesty International, Israeli authorities are using advanced surveillance technologies to entrench a new 'automated apartheid').[17]

The case against the Israeli government is interesting because it has also generated a coalition of dozens of activist groups working to expose the state's spyware industry. Groups like the Palestinian Boycott, Divestment and Sanctions movement are calling for a ban of this software.[18] But, as has been well documented, the danger is also global. Pegasus software – which is sold by the NSO Group under the protection of the Israeli state – has been used to monitor and target citizens in at least thirty-four countries including France, Germany, India, Mexico and Spain. The software has resulted in what Amnesty International calls massive human rights violations, including the harassment and murder of journalists and activists.[19]

Other important alliances are emerging between activists resisting data colonialism and those working in areas like labour, consumer protection and the environment. Take for example the emerging global coalition of individuals (from Chile, Ireland, the Netherlands and elsewhere) who are challenging how data centres use vast amounts of energy and contribute to global warming.[20] Or take the project Our Data Bodies whose researchers have worked closely with marginalised US neighbourhoods in North Carolina, Michigan and California to understand how data practices impact on them differentially and to support the evolution of alternatives.[21]

How Indigenous Communities Are Resisting

First Nations understand colonialism like no other community, having experienced and resisted it directly for centuries. There are many aspects to this resistance, including fighting for food and water sovereignty. But data is also becoming part of the struggle. An Indigenous Data Sovereignty movement has materialised to ensure that indigenous people maintain autonomy over their data resources. Indigenous people seek to control the data they generate, how it is used, and even what it means for their communities now and in the future.

Instead of rejecting digital technologies altogether, indigenous communities are thinking carefully about how they can be appropriated to support shared goals and, as indigenous Mixe theorist Yásnaya Elena Aguilar Gil puts it, 'repurposed for resistance'.[22] These communities are building their own telephone and internet infrastructures (just like they have been doing with radio for a long time), creating their own video streaming platforms to tell stories in their own narrative structures, writing their own Wikipedia entries, and developing their own apps and web browsers in their own languages.[23]

The issue of indigenous languages is particularly important to these communities, since it is estimated that between 50 and 90 per cent of languages will be endangered or dead by the year 2100.[24] Yes, Big Tech offers its own 'solution' to language disappearance through datafication, digitising indigenous voices to build speech recognition apps that can be used to keep those languages alive, but this only replicates colonial structures of dependence. In indigenous communities like the Māori in Aotearoa (otherwise known as New Zealand), people are rejecting that model. To save the indigenous language *te reo*, community activists Peter-Lucas Jones and Keoni Mahelona have developed their own AI language tools, and developed methods to collect and manage language data in a manner that allows the community to retain control and ownership of the process. When describing why they opted to build their own digital hosting platforms, instead of simply uploading everything to a Big Tech platform, their answer was very clear: 'Our data would [that way] be used by the very people that beat that language out of our mouths to sell it back to us as a service . . . It's just like taking our land and selling it back to us.'[25] In a contrasting case, volunteers on the Iran–Iraq border contributed hundreds of hours of labour to ensure that Sorani Kurdish was available in Google Translate. The company paid them nothing and now owns the results of their work.[26]

We come here to a fundamental issue: the desire for autonomy. More and more groups across the world want to use digital technologies and data to ensure the continuity of indigenous, marginalised, minoritised or low-resource languages, but in ways that they control. The survival of their language means the survival of their culture, which means securing the ability to use it safely and on their own terms in an online world whose technologies they also want to control.[27]

How Everybody Can Resist

The movement for refusing data colonialism goes still wider. It includes cultural workers: artists, filmmakers, novelists and journalists. It also includes students and teachers, politicians and constituents, and everybody in non-profit organisations working to ensure the right to encryption or the right to repair. Resistance can be large-scale (boycotting a Big Tech platform) or small-scale (refusing to accept the HTTP cookies when visiting a website (which apparently only 0.5 per cent of users do).[28] Even putting down our phones for a few hours might be an act of defiance.

It is easy to make fun of the influencer announcing to their social media followers that they are doing a weekend digital detox. But we are seeing ever more examples of individuals who are quitting these exploitative technologies.[29] And businesses too: not everyone will be able to take the financial risks that the CEO of Lush did, who said he was willing to lose US $13 million in revenue as a result of closing their social media accounts,[30] but it is inspiring nonetheless.

More accessible forms of resistance do exist. Take the teenagers in New York who have formed a Luddite Club and exchanged their smartphones for flip phones or no phones at all. One of them proclaimed: 'You post something on social media, you don't get enough likes, then you don't feel good about yourself. That shouldn't have to happen to anyone'.[31]

While the actions of any single individual or group are generally invisible, collectively they demonstrate that the movement to resist data extractivism is going on in many places and taking many forms.[32] Inspired by the thinkers and activists we learned about in the last chapter, we can reclaim the moral force of resistance, develop new ways to link past and present, and rethink the project of humanity. A certain mix of outrage, curiosity and passion for justice may be required to join this anti-colonial movement. But since everyone is affected by data colonialism, resistance can come from people in all walks of life and all sorts of places. One common denominator is crucial and it is the greatest weapon against colonialism that has ever been designed. The Luddite Club, and everyone else resisting data colonialism, already seem to possess it in abundance: imagination!

Radically Reimagining How We Use Data

'Decolonisation' can mean many things, not all of them radical or transformative. But if we're serious about it,[33] decolonisation means the incredibly difficult work of ceasing to see the world from the coloniser's point of view, and then imagining and implementing alternatives. While moving beyond the legacy of colonialism in every aspect of social life is a task beyond the scope of this book, let's think what this would mean in relation to data.

Decolonising data means dismantling extractivist models for collecting data, rejecting profit-driven views of our life as a territory ripe for dispossession, and turning data into a collectively owned tool for transforming the world in positive ways. But how can we do this?

A good starting point is to think about what data colonialism means for our personal space and time: can we imagine decolonising the space and time we give up to platforms and processes that extract data from us? The space colonised by the familiar data-extracting technologies – surveillance cameras, fitness trackers, Smart appliances, digital assistants, data-extracting chips under our skin – that turned *our* space, the space of our lives, into a territory of continuous extraction for profit. And the time colonised by these same data technologies, as we interact with them, and supply them with information about ourselves; a time filled with alerts, notifications, updates and reminders that has become less and less our time, and more and more the time we must surrender in service to the coloniser's agenda.

To decolonise our space and time may involve learning the art of what artist and writer Jenny Odell calls 'doing nothing'. This is not a defence of procrastination, but something much bolder. It means disengaging from the attention economy, the economy that revolves around getting us to spend more time online in order to extract more data from us, so that we can re-engage with something else. 'That "something else" is nothing less than time and space', Odell writes: decluttering our attention, bringing our focus back to what surrounds us, and how we spend our time in it.[34] Re-engaging with nature, with each other, and with the challenges of the world. Having said that, we should recognise that not everyone can afford to disengage from platforms, particularly those in precarious economic situations who might depend on the gig economy to make a living. But the dominant power of Big Tech will not be defeated unless its decolonisation starts somewhere.

New feminist struggles are a huge inspiration for this imaginative politics of resistance. One powerful thinker is the Argentinian Veronica Gago. For her, the problem is always extractivism: treating women's bodies and labour as 'zones of extractivism' open to seizure by males in what she calls the 'colonization of new territories'.[35] Gago insists that it is the space and time of daily life that is the first frontier to be defended, since 'domestication and colonization are inseparable'.[36] Which means that liberation is possible only when we confront how the system dehumanises people in a daily and intimate manner. This opens up some very difficult questions: Can we start to unthink common notions of private property? Can we open up to some ideas of collective property? When thinking about our data, these questions could be the start of a longer imaginative journey in which feminism plays an important part in resisting data colonialism.

Earlier critiques of modern technology can help us, too. Take the theologian, philosopher and social critic Ivan Illich, who was born in Austria but lived much of his life in Mexico. In his 1973

book *Tools for Conviviality*, Illich reflected on how our tools tend
to detach us from our real goals, in the process creating 'useful
things for useless people'.[37] Illich suspected that when the 'useful'
technologies of the industrial age offer us a supposedly easier and
more convenient life, they encourage us to surrender a degree of
control and autonomy over life itself. Consider WhatsApp or
WeChat: it might make it 'easier' to communicate with friends and
family, but thereby converts everyday communication into an input
of a profit-generating process. Put bluntly, in the long run a life full
of 'smart' devices is really not smart at all.

The answer, Illich proposes, is a society of responsible tools, a
convivial society in which humans control their technologies, not
the other way around. What makes this society convivial is that its
'technologies serve politically interrelated individuals' in ways that
promote 'autonomous and creative intercourse among persons'.[38]
Conviviality, as Illich uses it, is an evocative word. It suggests not
just a warm atmosphere, a *craic* as the Irish term puts it, but a tech-
nopolitical project in which individual freedom gets realised only
through the interdependence between people. Far from being
against technology, Illich saw that our interdependence can be facilitated by our technologies, but only if we put careful limits on our
tools so that they work *with* us, rather than govern us in ways that
benefit multinational corporations.

Illich's ideas link to radical feminist thinking, to indigenous
thinking, and to recent proposals for disconnection. While his writing is half a century old, it still captures the need for new forms of
resistance. Decolonising data is not about renouncing data per se,
but about embracing models that produce convivial data: data that
can help us relate to each other responsibly and better, and that can
help us protect each other and the environment. And the core, once
again, is our human creativity and imagination. Trust in this, and
we can have hope that convivial data, data that serves actual communities, will become possible.

To move in this direction requires sustained effort. Projects in Data Science for Social Good (DSSGx) are becoming trendy crowd-sourcing exercises sponsored by Ivy League institutions, and even the G8 have announced that they will push their governments towards making data open by default.[39] But generating data and making it accessible to the public is not enough. After all, we already have plenty of data on global warming or the effects of racism, which has obviously not been enough to convince some individuals, corporations and governments to change. The larger and more difficult goal of transforming data into better social relations is what's at stake in 'convivial data'. But already there are examples of just this, such as initiatives that use local data to address structural racism in communities.[40] Others, such as the collective Data Against Feminicide (datoscontrafeminicidio.net), train organisations to standardise data collection practices, develop data analysis tools, and think about what it means to serve as stewards of that data.[41]

Precisely because decolonising data is a project for living differently with data, it must encompass not one answer, but many. It must formulate solutions that are not only technological but social, political, regulatory, cultural, scientific and educational. And it must connect to struggles that seemingly have nothing to do with data, but that in reality are part of the same underlying struggle for justice and dignity.

Some people might reject this search for alternatives out of hand, as evidence of a desire to do away with Western reason, even Western civilisation altogether. But, as we saw when we considered Quijano's work, that is too hasty. If we learn anything from the history of colonialism, it is that concepts like reason, pluralism and autonomy are not constructs to which only the West, or Western models of society, can lay claim. In fact, Western thinking 'borrowed' (to put it euphemistically) many of these concepts from non-Western traditions.[42] So it's not a question of doing away with rationality; it's a question of doing away with the Western claim (echoed by Big Tech) of exclusivity over rationality, diversity and autonomy, as if only Western models of those things could exist.

Once we begin to multiply and apply the alternative rationalities and alternative ways of producing and applying data described above, the process of decolonising data begins in earnest. We are already seeing communities take control of their data, take control of the process of putting limits on data technologies (following Illich), and reimagine what data is and what it's for. These communities are asking, for instance: 'Can AI be decolonised, or is there something intrinsically colonial about the way in which its algorithms are trained through extractivism?' No one has definite answers yet, but the conversations have started. Honest discussions about our own role and participation in data colonialism have also started. These dialogues are oriented towards values such as autonomy, care, community and reason.

What matters above all is to interpret these values not in a narrow way that feeds power and domination, but in an inclusive and pluralistic way. There's still great power in the vision of the Zapatistas' land resistance movement in the southern Mexican state of Chiapas 30 years ago, who proposed 'a world in which many worlds fit'. Data, if produced and used in the right way, can fit with this plural vision of the world, and create new types of data territories, where data is gathered, processed and managed according to the goals and norms that real communities design and approve.

Introducing the Playbook

One thing is clear: single-track approaches to solving data colonialism are not going to work. We can try to quit this or that platform, pass this or that law . . . but by themselves, isolated actions are not going to have much of an impact because of the complexity and persistence of the problem – which is, after all, part of a larger problem with a long history! We need to be able to think and act more holistically. The resisters and thinkers that have inspired this book have all done just that.

The activist organisation Mijente proposed a framework that can help. This framework was adapted from Latin American activists such as the Pobladores en Lucha (Residents in Struggle) movement in Chile.[43] Taking inspiration from this model, collective resistance on the largest scale always means tackling data colonialism at three levels simultaneously: by working within the system of data colonialism, by working against the system, and by working beyond the system.

Play #1: Working within the System

Having read about the multiple abuses perpetrated by data colonialism's corporations, and the nation-states that benefit from them, your first reaction might be to abandon any kind of collaboration with representatives from the private and public sectors. That type of response has its uses, which we will discuss later. But walking away from this kind of work risks strengthening the hand of governments and corporations who might not have our best interest in mind. What we mean by working with or within the system, then, is that we must go on demanding our rights from states and corporations, even if they are unwilling or unable to guarantee them. The moment we cease to demand those rights, we give governments and corporations permission to do as they please, and to forgo any kind of obligation to us. Even if we know the serious limits to any response we expect to get – if our grievances are even heard at all – we must go on exercising our rights as citizens and consumers, whether at the local, national or international level.

What kind of pressure can we exert within the system to decolonise data? We need to push for better and more regulation of the market. Attempts so far to rein in Big Tech have been slow.

The European Union has made important strides in formulating frameworks for what a well-regulated Social Quantification Sector might, in part, look like. Recent European legislative changes are too complex to discuss in detail here. But without question the GDPR (General Data Protection Regulation), which came into force in 2018, laid down a basic challenge to the idea that Big Tech market power should grow without restraint by insisting on the human rights implications of 'the processing of personal data'.[44] New legislation such as the Digital Services Act and the Digital Markets Act goes further and regulates the services that platforms provide and the asymmetrical market structures from which they benefit. Meanwhile the EU Commission has imposed a record fine on Meta (€1.2 billion) and is considering a ban on facial recognition in public places.[45]

Many legal commentators worry that EU legislation relies too heavily on the principle of consent: giving consumers the opportunity to consent, or not, to whatever Big Tech wants to do with their data. As noted in earlier chapters, data gets extracted in many circumstances where real consent is impossible, since force plays a greater role. And even the best legislation is only as good as its enforcement by courts and regulators. For example, Amazon recently avoided a multi-billion-dollar fine for monopolistic behaviour by agreeing to make some changes to its shopping website, changes which it was already planning to do.[46] And the new EU legislation, even when it comes fully into force, may face court challenges by Big Tech giants. Broader doubts also remain about whether EU legislators really want to halt the expansion of Big Tech's data extraction or just to ensure that data markets operate more effectively in accordance with narrowly defined privacy rights.[47] If so, the expansion of data territories will not be halted.

Worse, in other places like the United States and the United Kingdom, significant regulation has been delayed. These are the places where laissez-faire capitalism promotes the disastrous illusion that it is best to let corporations regulate themselves, and where corporations employ powerful lobbies to give money to politicians in return for favourable deals that allow them to continue to do as they please. This is exactly what worries many NHS data scientists as they face the prospect of a long-term dependence on a corporation like Palantir, for example.[48]

Rushed legislation that simply attempts to prove that governments are doing something can be just as dangerous, as in the case of the Kids Online Safety Act (KOSA), introduced to the US Congress in 2022. A coalition of over ninety civic organisations critiqued the bill on the grounds that, while claiming to protect children's privacy, it would end up exposing them to more invasive tracking, filtering and monitoring tools.[49] The act was rejected but has been re-introduced in 2023 with minor modifications.

Change within the system is always going to be slow and complicated. But it is possible. Anti-trust regulation is one area that looks promising, even in the US. People like Lina Khan, current chairwoman of the US Federal Trade Commission, are taking a very close look at the question of whether companies like Amazon and Meta are too big, and whether their monopolistic practices serve consumers well. In August 2022, Khan issued an 'advance notice of proposed rulemaking' about the damages caused by commercial surveillance. This has generated a large number of responses from business and civil society. A debate is under way about how to 'change the paradigm' that has accepted corporate surveillance for so long.[50] But it remains unclear how much the FTC will be allowed to do if a deeply conservative Supreme Court and a potential Republican government in 2024 have their way. Big Tech nonetheless fears these proposals, which is why they have attempted to force Khan to recuse herself from such investigations, as if her commitments to address Big Tech's problems make her unqualified to implement them.[51] Wherever we look, this is going to be a long war.

A different and potentially bolder approach to regulation would involve the nationalisation of data by countries, a proposal made particularly by countries in the Global South. This would involve declaring data a national resource, as some countries have done in the past with oil or other natural resources and industries. If data is nationalised, governments could demand that foreign tech companies pay a tax for using this resource, so that its exploitation benefits the citizens of that country more directly. But we need to keep in mind that nationalisation would not necessarily stop extractivism, and that as an anti-colonial policy, it has a history of mixed results. In response, Argentine scholars Leonardo Fabián Sai and Sofía Beatriz Scasserra have proposed a new framework for thinking about data as a raw material whose use can be regulated by the government within a wider re-industrialising process based on shared access to data as a common social resource. But further work will be needed to ensure that these models avoid social exploitation when implemented.[52]

The motivation behind all of these different regulatory models is a belief that the status quo is no longer working, and that we need broader and more ambitious regulatory frameworks against data colonialism. Meanwhile the corporate arguments against regulation live on, including the idea that, because data technologies are more complex than regular technologies, it is more difficult to regulate their predicted impacts. But surely this should be a reason to demand more regulation, not less. Panicked claims from the private sector that regulation will kill innovation should not be taken seriously. In any case, if innovation means discrimination and exploitation, as data colonialism demonstrably does, why should we as citizens accept this?

Even if progress will be difficult, we must use what opportunities we can to work within the system. Here are our proposals.

Ideas for Acting within the System

- **Start a group within your company to protest inequitable data practices.** As we have seen, some employees are becoming more vocal about protesting the way in which their work is being used for ends which they consider unethical or exploitative. This comes at the risk of being fired or penalised, of course, but in some cases workers have been reinstated following public outcry. The example of Google workers forcing the company to suspend its Pentagon contract for Project Maven is noteworthy.[53] Most corporations, especially those who see themselves as forces for good in the world, remain sensitive to their public image.

- **Work with your local governments (city, municipality, etc.) to enact social change.** As activists know well, local governments might be more accessible to ordinary citizens seeking to change things than larger governance bodies.[54] This means that trying to pass a ban on facial recognition or predictive policing algorithms might be a more attainable goal at the local rather than at the federal/national level. In addition, cities and towns (even small ones) have purchasing power, and they can decide how to spend their money. Getting a city to reject a contract with a questionable provider can send a strong message. In one case, citizens in San Diego worked with their representatives in City Council to rein in the Smart Streetlights programme, which attempted to introduce sensors and cameras throughout the city.[55]

- **Work with other local organisations to effect social change.** The strategies described above can be replicated in other areas of administration, not just government. These can include faculty and student university assemblies, parent–teacher boards, church and charity boards, social clubs, advisory boards, and so on. One approach might involve working with bodies like these to change the allocation of their financial investments in more socially responsible ways (that is, away from tech corporations that are involved in extractivism and abuse). As the Boycott, Divestment and Sanctions movement has demonstrated, this can be a very effective strategy.

- **Raise awareness through a literacy campaign.** A well-informed public is the basis of change from within the system. Any efforts to share what you have learned in this book, as well as the results of your independent research and experience, can be instrumental to changing others' opinions and perceptions. There are plenty of outlets beyond the traditional letter to the editor, including social media posts, podcasts, book clubs, classes or activities within classes, or simply discussions with friends and family. If we're talking about working with the system, then feel free to *use* the system's resources against itself! This is what resistance movements have done in the past.

- **Elect (and help get elected) representatives with the right agendas.** Many people have become disillusioned with politicians who seem to only be interested in fuelling cultural divisions. But getting politicians to listen and take action is still essential work, as is voting itself. Although the results are often uneven, there are, as we've seen in the many examples above, signs that some politicians are listening.

- **Support progressive taxation.** Tech companies don't pay enough taxes. For instance, from 2018 to 2021 Amazon paid about a quarter of the federal corporate tax rate in the US, even though its profits increased 220 per cent during the pandemic.[56] We should support campaigns to demand that corporations that profit from our data pay more taxes (recent efforts by the Organisation for Economic Cooperation and Development to do this were blocked by the US and Silicon Valley, but other attempts at the state and nation level continue[57]).

- **Support the implementation of Nobel Peace Prize Laureates Maria Ressa and Dmitry Muratov's 10-Point Plan to Address the Information Crisis.** This plan calls on democratic governments to protect citizens' privacy with data protection measures, condemn attacks on journalists and safeguard their safety, ban surveillance advertising, challenge corporate lobbying, and enforce the emerging laws that make these things possible.[58] One organisation promoting this plan is People vs. Big Tech (peoplevsbig.tech).

- **Demand and participate in Algorithmic Impact Assessments (AIA).** Impact assessments are commonly used to measure the economic, environmental and social ramification of projects implemented by state or private actors. So why not in the data sphere? AIA are evidence-based studies that can empower the public with clear information about the automated decision systems that impact their lives. They can also help public agencies and supporting experts to develop the expertise to review systems with discriminatory impacts (in areas like welfare distribution, predictive policing, energy use allocation, labour rights and access to education).[59] Interventions of this kind need to become more commonplace whenever data technologies are being implemented. For an example of what this might look like in practice, see the user guide prepared by the Ada Lovelace Institute to explore the application of AIA in healthcare.[60]

Play #2: Working against the System

While working within the system is a must for a fully engaged activist, and can have productive outcomes, it is also a limited strategy.

There is only so much change we can expect from the parties that are largely to blame for the current situation in which we find ourselves. Colonialism was always a partnership between the private and public sectors: corporations supplied the logistics and the management resources (the East India Company is a good example), but governments and, earlier, the Church supplied the legal and theological justifications. Today, the partnership continues, and while some governments do occasionally manage to put the interests of citizens above those of corporations, we must remember the obvious and not-so-obvious ways in which these colonial partnerships have retained their influence.

In short, working within the system might make life a bit more tolerable for the colonised, but it is unlikely to fundamentally challenge the status quo. For that, we need to develop new political tools that issue a big resounding 'no' to data colonialism's agendas.

This tension between working within and against the system brings us back to the issue of sovereignty. Can we become truly autonomous if we still depend on the master's tools? We are recalling here the famous insistence of Audre Lorde (a Black American writer) that the master's tools will never dismantle the master's house.[61]

Changing one set of oppressive tools for another that doesn't guarantee our autonomy is not enough. Is it better for people to use the web browser Opera instead of Chrome, if the former is now owned by a consortium of companies from China, and the latter by Google? Does it benefit people in Latin America to use Colombian-based app Rappi instead of US-based apps DoorDash or TaskRabbit, if all those apps employ similar algorithms to exploit gig workers doing errands? Will the African super-app Yassir solve the issues of data colonialism, just because it is run out of Africa?

Sometimes the concept of state sovereignty (of the state as the ultimate arbiter of internet policies) has been co-opted by parties and individuals for political gain, creating new tools of surveillance and oppression. For instance, India's draft Telecommunications Bill while framed as a corrective to the colonial-era Telegraph Act, proposes to outlaw encryption in apps like WhatsApp, Telegram and Signal (giving the state authorities the ability to monitor private messages), as well as give the government the power to suspend internet access in times of crisis.[62]

In short, choosing an internal coloniser over an external one does not really improve the plight of the colonial subject.[63]

At bottom, it is less about ownership than about purpose. Repurposing the master's tools so that they can work against the system means actually starting to say 'No': rejecting the civilising stories that tell us that unchecked data extraction is acceptable and normal. Saying 'No' can operate at the individual level, as we make personal decisions about what we can accept or reject in order to protect our basic autonomy.[64] Sometimes saying 'No' means acknowledging that we have no realistic alternative in the short-term but to accept some aspects of data colonialism individually, even if we reject their principles. But saying 'No' also means being open to joining wider collaborative projects that look to undermine the system in the longer run.

There have been various proposals, put forth by well-known figures such as Tim Berners-Lee (inventor of the World Wide Web) and Jaron Lanier (one of the founders of the virtual reality field), to 'fix the internet' by developing platforms that would allow people to be freely able to transfer their data, or to get paid for the data they produce.[65] While these proposals raise important questions, the problem is that they tend to see the damage done by data colonialism as problems that have technological solutions, solutions that don't disturb the extractivist status quo. Being paid for 'your' data incorporates you and your data all the more effectively into the system, incentivising us to generate data and move it more fluently between data-extracting platforms.

Other proposals focus on the adoption of Web3, which will sup-posedly break up Big Tech power. Without going into mind-boggling technical details, Web3 is about using blockchain (the technology behind cryptocurrencies like Bitcoin) to re-structure the internet. Whereas Web1 was the 'read' web (users could only read static con-tent), and Web2 is the 'read-write' web (users can create and modify content, but the platforms for doing so have become monopolies), Web3 will be the 'read-write-own' web, which takes advantage of encryption, decentralisation and distributed decision-making tech-nologies to allow users to protect their identities, own their data, bypass censorship, and allow them to directly invest in products with-out intermediaries like banks and corporations. It remains to be seen how this model could – in actuality, not just in theory – shatter monopolies and re-structure the economy in favour of the average user. But even if it can, it is already clear that the problems associated with extractivism will not magically go away. As critics have already pointed out, Web3 is rife with financial speculation (remember the non-fungible token craze, and the number of people who lost money on it?), it's an expensive and not very practical technology, its anonymity encourages abuse and harassment, and it consumes a lot of energy, which is bad for the environment.[66]

These 'solutions' remind us again that the problem of extractivism is at root social and political, not just technological, and our solu-tions need to be social and political also. In response, US tech scholar Ben Tarnoff has proposed that we deprivatise the internet, that is, shift it from private to public ownership and control, as is only fit-ting for a resource that really does support our common life. That's why the postal service, education, transportation or healthcare in many countries are public services. By his own admission, Tarnoff's ideas remain at an early stage, but by transforming how we think about the internet, they provide a much better starting point than pure market-based solutions.[67] An alternative that has already had some practical success is offered by the Platform Cooperativism

Consortium (platform.coop), started by Trebor Scholz. While co-ops have been around for centuries, this consortium has built a global movement revolving around democratic governance, co-design, and open technologies.[68] Rather than simply re-inventing Uber or focussing on tech-centric, venture-capital-funded solutions that benefit shareholders, the consortium helps communities form cooperatives that provide fair and dignified working conditions for their members. The goal is to build something that provides broadly the same services as commercial platforms, but owned on a more collective basis with a less extractive business model. At this point, rebuilding platforms to tame the colonial expansion of Big Tech connects with other movements that aim to socialise capitalism.[69]

All of the above proposals, even with their limitations, can provide us with parts of the puzzle. Some of the tools and models to oppose data colonialism are, in other words, already here. But we need to bring them together in a more cohesive way, which means collaboration between players who currently work apart from each other. How exactly do we create a culture of resistance to data colonialism across groups with different goals and contexts? Any real progress towards that requires the following three steps.[70]

First, identify common principles. Part of the reason, we think, the data colonialism framework has inspired so many people is because it provides us with a set of shared concepts to understand the problem, concepts based on histories that unite the colonised throughout the world, and that suggest principles to resist it. The second step is to build solidarity as a concrete way to help each other bear the costs of resisting the system. Solidarity can be one of those aspirational words that no one is exactly sure how to make real. Ironically, data colonisers are very good at supporting each other to maintain the status quo, while discouraging the rest of us from believing we can act together to dismantle it. Solidarity, when backed by a strong sense of shared responsibility, can help us name our common enemy and our common goal. The last step is to create political, ecological and economic literacy. When working against a system – any system – this kind of social knowledge is particularly important. The good news is that the knowledge about how to resist data colonialism is actually here already in some form or other. It comes to the surface in people's anger when a scandal such as Cambridge Analytica erupts, only to sink beneath the surface again as normal data-extracting life resumes. Our goal in this book has been to catalyse those individual intuitions about what's wrong with data colonialism into something broader and more durable. Let's start sharing our private knowledge that something is wrong with data extractivism and start building practical cultures of resistance. If we stop thinking of our critical intuitions as individual quirks, we can start to see them as part of an emerging shared knowledge, a collective awareness that data colonialism and the extractivism on which it is based are just wrong.

Ideas for Acting against the System

- **Examine your own role in data colonialism.** Self-critique is useful, but perhaps even more important is the ability to hear critiques of us by others. Examine your own role in replicating patterns of power in daily life, within the structures you inhabit. Is your own voice dominant, or silent? The goal is to avoid either of the two extremes: complete disavowal that we are part of the problem, and complete paralysis because we know we *are* part of the problem.

- **Expose bias in algorithms and data technologies.** If you work for an organisation that uses algorithmic decision systems, and you notice those systems are biased or unfair, consider publishing (with others who may feel like you) an internal exposé, blowing the whistle, or going to the media, if those represent acceptable risks in your situation. Or support the work of organisations that are already doing investigations in this area such as Algorithm Watch in Germany or The Markup in the US.[71] You can also contribute to the important work being done at organisations like the Algorithmic Justice League (https://www.ajl.org), founded by Joy Buolamwini, to promote awareness and social change through research and art.

- **Change society by changing the curriculum.** If you are a student, you can demand that your schools do a better job preparing their graduates to think critically about technology. In fact, you can get involved in the process of redesigning the curriculum on how to do it! We can fundamentally challenge the system by changing the way schools prepare and shape the minds of the next generation of those who design data technologies, as well as those who think about and critique their application. There may already be teachers at your school doing this, and they would love to partner with you. And, outside educational institutions, let's challenge what Joseph Turow, in his critique of marketing surveillance, calls the 'hidden curriculum' of the corporate surveillance world.[72] *Don't* install Alexa in your home and, even more importantly, *don't* let your children grow up thinking Alexa is normal.

- **Support gig workers (or any workers) who are fighting to unionise or are demanding their rights.** The labour movement is experiencing a renaissance, as we described above, and we need to support all workers organising to unionise, and all workers demanding labour law reforms. Check out what they are up to (Wikipedia has a useful page listing global unionisation efforts in the tech sector[73]) and signal your support.

- **Share the stories of individual victims, and work with others to defend the most vulnerable.** Human beings respond to narratives, and it is very effective to capture the essence of a problem like data colonialism by focussing on an individual's story (in a non-exploitative way, of course). Good examples of this are the work of journalist Karen Hao (already cited previously in this book), Data & Society's project Parables of AI in/from the Majority World,[74] and the Domestic Code project, which collects stories of women in the cleaning and care industries working through gig economy platforms.[75] If these stories can be shared without using extractivist social media, that's great. But where social media does allow you to reach the audience you need, use it! No movement for change has ever been able to operate without using the tools to hand, including those of the system that it is fighting.

- **Do something practical to support organisations and research institutes working in these areas.** There are now hundreds of organisations across the globe working on issues including online privacy, consumer rights, digital justice, technology and democracy, platform cooperativism, internet governance, and so on. We have been trying to name some along the way. While many of these organisations work within the system, others take a more oppositional stance. We found a good list of them at consentofthenetworked. com/get-involved. There is also important research being conducted at universities in the Global North and South to expose the dangers of data extractivism. To give just three examples: Observatório Educação Vigiada is a project by researchers in Universidade Federal do Pará, Brazil, that collects data about the platformisation of public education in South America.[76] Meanwhile the University of Cambridge's Minderoo Centre for Technology and Democracy has created an audit tool to monitor compliance with current legal guidelines, and has shown that UK police have failed to meet legal and ethical standards in the use of facial recognition.[77] Finally we would mention the research on data justice as it affects law enforcement, welfare systems, the workplace and many other areas done by Cardiff University's Data Justice Lab.[78] This kind of research needs to be supported and promoted.

Play #3: Working beyond the System

Working within the system means to accept – by necessity, if not by conviction – the worldview that makes such a system possible, with the hope of transforming it internally. But working against the system is equally bound by that worldview, even if in a more subtle way. For by defining ourselves in complete opposition to something, we are still letting that something define us, as Umberto Eco observed.[79]

Which is why the third strategy for fighting data colonialism involves not working from within, not even working against data colonialism, but attempting to imagine a different worldview altogether, a worldview not conditioned or circumscribed by the very thing we are trying to dismantle. This is where creativity is most needed.

Can our imagination ever be completely free of the social context in which we live? Probably not. But human imagination always finds fractures in the social or economic system from where it is possible to 'unthink' that system, spaces and practices where, even if only briefly, different ways of doing and thinking can emerge.[80]

Here are some inspiring examples of how local communities are forging these spaces and practices, developing alternative models of data sovereignty. We learned about some of these projects from our colleagues in the Tierra Común network, Paola Ricaurte and Rafael Grohmann (more on Tierra Común shortly).[81] One is the GeoComunes data journalism collective (in Mexico) that helps small communities use cartographic data to map the ways in which environmental degradation and privatisation impacts them. There is also InfoAmazonia (in Brazil), an independent media organisation that uses geolocalised data (data about specific points in space) to report stories about the endangered rainforest region. Other examples include feminist or trans hacker cooperatives such as MariaLab (Brazil), whose goal is to 'bring technology to feminist spaces and feminism to technological spaces', and Alternativa Laboral Trans

(Argentina), a non-profit collective (mentioned earlier) founded by trans people that provides web design and education.[82] There are many organisations in Latin America involved in digital rights issues, including Datysoc in Uruguay (datysoc.org), Hiperderecho in Peru (hiperderecho.org), InternetLab in Brazil (internetlab.org. br/en), Karisma in Colombia (karisma.org.co), Openlab in Ecuador (openlab.ec), and Vía Libre in Argentina (vialibre.org.ar). Two good examples of organisations doing digital literacy work are Sulá Batsú in Costa Rica (sulabatsu.com) and Digital Empowerment Foundation in India (defindia.org). There are too many other organisations doing remarkable work around the world for us to list here, including in Africa, Asia and Oceania. But we also want to mention the important work that activist-artist collectives such as Sursiendo in Mexico (sursiendo.org) are doing to promote the sustainable self-governing of digital communities by working within the framework of 'permaculture'.

In the context of this struggle, the projects we have mentioned have one thing in common: they reject the norms of data colonialism and insist instead that data can be collected, processed and analysed in accordance with community principles, not commercial ones. It really is possible to create a space where we develop our own tools and models for reconfiguring conviviality, as Illich would say, redefining what it means to connect to other human beings through data. From these fresh starting points, other ways of thinking about what data is and what it is for *can* emerge. And, as we do this, we may start to discover what it means to assemble a viable anti-colonial system for socially managing data at home, at work, at school, and in the public sphere.

This brings to mind the work of Brazilian educator Paulo Freire, who argued that the struggle for liberation is a struggle that re-humanises not just the oppressed, but also the oppressor, whose treatment of the oppressed as objects – as mere data, we could say – ends up dehumanising them too. Freire saw the process of education, of becoming literate, as one that entailed the act of renaming the world in order to actively change it.[83] A substantive critique of data colonialism can help us not only to name the data extractivist system and its many parts for what they really are, but also to rename and recreate the world of data so that it becomes a genuinely social resource. These deeper challenges to data colonialism could, in the long run, have transformative impacts on society's and capitalism's wider operations.

Ideas for Working beyond the System

- **Embrace the power of critical thinking.** As authors, we have been told that the message of data colonialism is too depressing. But we believe, as Italian intellectual Antonio Gramsci suggested, that confronting the world's problems requires a pessimism of the intellect, along with an optimism of the will.[84] We take inspiration also from feminist thinker Sara Ahmed's figure of the 'feminist killjoy' who, yes, undermines people's sense of easy comfort and convenience by 'recognising inequalities as existing'.[85] While facing the enormity of five centuries of colonial oppression, let us counter the risk of hopelessness with the power of individual thought, collective imagination and collective action. That is the surest way of moving beyond the current system.
- **Help each other become less dependent on data-extracting platforms in our daily life.** While it might not be possible for most of us to simply quit all of these platforms in one fell swoop, or to convince our families and friends to join us in doing so, we must remain vigilant for unexpected opportunities. For instance, the scandals that plagued Twitter after the takeover by Elon Musk provided an opportunity for hundreds of thousands to leave the platform and explore non-extractive alternatives, such as Mastodon. Unfortunately, after an initial

bump in membership, Mastodon experienced a sharp decrease in active memberships, demonstrating how difficult it is to replace corporate platforms. Nonetheless, it remains an interesting model for how, in principle, extractivist platforms might be built differently, providing inspiration for further thinking on how to build the social resources that can sustain a different type of network.

■ **Support or join initiatives engaged in building new ways of gathering the data we need, protecting it and using it for good ends.** As we've stated before, decolonising data does not mean rejecting data in all its forms. It is possible to generate the data we need to understand the world and make it a better place, and the organisations and projects we've mentioned in this book deserve our support to do so. Each one of us is entitled to play a role in deciding how data from our neighbourhoods, our schools, and our hospitals are used.[86] When we, not corporations, get to define what data we need to build new worlds, we define new principles for producing social knowledge guided by social values, not commercial ones.

These action items are somewhat more abstract than the previous ones, and that's for a reason. We do not believe there is a universal checklist of 'correct' strategies to work outside the system of data colonialism. Each community must define them according to its context, while working to contribute to a shared goal.

Speaking of communities, we should mention Tierra Común, the network that we co-founded with Paola Ricaurte, that has brought together nearly one hundred activists, scholars and educators (mostly from Latin America) interested in decolonising data. It is not possible to list all of our members here by name,[87] but their efforts encompass areas as diverse as decolonial computing, indigenous data sovereignty, and community and indigenous communication. When it comes to gender and power issues, Tierra

Común members are working on feminist technology, online gender violence, feminicide data, algorithmic discrimination, political and ethical ramifications of algorithmic media, and power differentials in datasets. They also look at the intersection of digital technology and local development, and the democratisation of knowledge. Some of them focus on digital labour and working conditions within platform economies, as well as the platformisation of public education. Others look at risk and cybersecurity, non-Western perspectives in AI governance, and human-centreed AI. There is interesting work being done in youth and technology, media consumption in Latin America, democratic and artistic approaches to programming and machine learning, and agile data storytelling and visualisation. And there are also those interested in justice in data-driven public policy design, design justice, connectivity and digital rights, and the geopolitics of digital rights (please see tierracomun. net for more information). This extensive list shows the rich work being done to challenge data colonialism within, without and outside the system.

We believe that, when applied to the data domain, decolonisation can become a narrative that unites us, a narrative that can help us redefine the border between what we want and don't want large-scale technologies to do for us. In short, a new narrative about how we can contribute to our collective human knowledge in non-extractive ways.

On a daily basis, as online platforms make our lives seemingly easier, or as we interact with a cool new AI gadget, it might look as if there is no urgent reason to change anything. Surely, we might think, what's happening with data is not as catastrophic as what is happening with the environment, for example. So what is at stake in resisting? To put things in perspective: the climate crisis we face, carried out with our participation but not necessarily with our informed consent, threatens our natural environment. The crisis of data colonialism, similarly carried out with our participation but not with our conscious approval, confronts us not with disappearing glaciers or rainforests (although data colonialism does come with significant environmental costs), but instead with impoverished social environments that are ruled by one exclusive purpose: the extraction of data in the pursuit of profit. That's a problem, because solving the environmental crisis – or any of the crises we face – requires social collaboration. If our social environments are controlled by states and corporations, they are at risk of being manipulated to serve their interests, not ours, and that will undermine the collective politics we need.

One lesson, above all, from the history of decolonial struggles is important here: that it is up to the colonised, not the coloniser, to take the first step in redefining our ways of being in the world. The forces behind data colonialism propose a world where they get to define how we relate to each other: they ask to determine at a deep level the conditions under which we live. But things needn't be this way, and they never have been until very recently.

As human beings, we have the capacity to name the world, rather than have it named for us. Let's name today's systems of data extraction for what they really are: the latest stage in a centuries-old project to rule the world in the interests of the very few. And let's imagine and name a different world, where data is something communities control for purposes that they themselves have chosen.

AND IF WE DON'T RESIST?

WHAT IF WE do nothing? What if we treat this book's argument as at best a diverting story and, at worst, a misleading slur on the fine goals of Big Tech businesses and platforms? After all, that is what most corporate and state powers the world over would like you to believe.

If we do nothing to resist data colonialism, it will certainly not implode under its own weight, even if that's the fate some commentators implausibly envisage for social media and surveillance capitalism. Because – as we hope we have shown – data colonialism is much larger than either of them. Data colonialism is an emerging social order that involves billions of people, an order that advances one stage further the colonial appropriations that are crucial to capitalism. So the actions of mere individuals will not affect it one way or the other.

But suppose we really do – all of us – nothing. What will happen then over the longer-term? For sure, the consequences will not be immediately obvious, because the power that drives data colonialism is disguised behind some very comforting alibis. The consequences of our failure to resist data colonialism will only unfold slowly and subtly, but in the medium-term, they will be devastating and irreversible. Let's examine why.

For sure, the enriching of AI through extracting ever more data from us may have its fun side. There will be ever more sophisticated recommendation systems that can predict our tastes with uncanny accuracy and give us constant access to what we think we want to consume (the undeclared but obvious objective behind the development of AI chatbots is to build more effective mechanisms for selling us stuff; that's how AI corporations really intend to make money).[1] Our games consoles and VR headsets will plug us into ever more nuanced interactions in online spaces, whether they are called the Metaverse or not. All our knowledge queries will feel less like an administrative task (typing something into a search box), and more like a conversation with a brilliant AI-powered friend who appears to know us better than we know ourselves. And, if we are to believe Open AI's chief technology officer, AGI itself will increasingly be integrated into everyday business applications, while AI chatbots become like another platform where we can search for information, while also 'buying groceries and booking restaurant tables'.[2] But these will not be the only changes from the datafication of everyday life.

We will see ever more areas of daily life transformed into impenetrable spaces we don't understand and have no control over, yet which are able to shape our chances of accessing crucial life resources (loans, education, healthcare, welfare and jobs). If data always discriminates, then weaving data and algorithms into the web of social decision-making will generate a biased system of unprecedented power.

There will be ever more types of work transformed into domains of absolute transparency for the manager, who will be able to track workers at every moment and predict their actions and manage their emotions, but increasing *opacity* for the worker, for whom management decisions will be shrouded in a black box called 'data'.

Governments will feel to citizens more and more like abstract interfaces of algorithmic calculation, opaque machines that watch, nudge and predict populations with ever less need to 'consult' them in person. The impacts on governments' legitimacy are unlikely to be positive.[3] And we will see the unrelenting growth of platforms like Amazon or Google or Tencent that become larger than governments in their resources and global reach, massively more powerful in terms of data-processing power and, as private corporations, unable and unwilling to offer even the semblance of democratic accountability.

Even worse, since corporations' power of large-scale data extraction will become increasingly normal – even for companies who don't own platforms or search engines – commercial forces will in general become significantly more powerful and substantially less accountable in people's lives. Whole sectors like agriculture, education and health that, until a decade ago, remained largely in the hands of practitioners will become shaped and managed by distant data-hungry corporations and platforms. New technological developments only glimpsed in their outline at present, such as AGI and the Metaverse, will intensify all the problems outlined in earlier chapters, converting everyday interactions and playfulness into ever richer data inputs. Even that most intimate interaction, a child receiving a bedtime story from her parent, may, within five years, be replaced by an interaction with an AI-powered cuddly toy that 'knows you go to which school . . . [and] who your friends are': that's the vision at least of toymaker Allan Wong, chief executive of Hong Kong-based VTech Holdings.[4]

Underlying all of this will be the continuous tracking of multiple dimensions of our lives – unless, perhaps, we are privileged members of data's colonial class. As a result, over time, we will come to feel less free, like Tracy (from Chapter One), who once felt she had 'nothing to hide', but over time came to realise that the fact that data was continuously being gathered about her could potentially undermine her sense of freedom. Less free, because once we cede control to external institutions over information about the intimate fabric of our lives, then we have started to cede control over who we can imagine we are.

And that possibility of imagining who we are is at the core of the idea of human freedom.

If the Chinese Communist Party's model for AI and all-powerful super-platforms represents the most extreme version of this transformation, there is lots of evidence that it differs only in degree from trends elsewhere in the world. All these trends are based on the same technologies and extractive logics. Technologies and logics that are, we saw, unsustainable in terms of the costs they impose on the physical environment.

Some – those who historically have throughout the history of colonialism been exploited – will feel this loss of freedom in their daily lives faster and more brutally than others. They will feel it in the decisions made against them that claim to be rational, but are based on harvesting of predictive data about their lives, decisions and processes that lie far from their control. They will feel it in injustices they understand and expect, but that are difficult to challenge. Over time, many of those who did not consider themselves disadvantaged by previous power relations will encounter new types of injustices delivered to them by data-driven systems whose impact no one seems to be accountable for.

In time, everyone outside the new colonial class of data hoarders will come to sense that something important and more general has been lost, even if, by then, our social world will have been so radically reorganised that we cannot quite remember what life without data extraction at every moment of our lives felt like. What will have been lost? Autonomy: each individual's right to exist free from external tracking and influence, the right to inhabit a space where we are just ourselves without relations to external corporations.

There are different ways of understanding autonomy, some highly individualistic, others (which we prefer) that are grounded in the mutual respect and social interchange of community life.[5] But however we understand that term, the idea that we must protect the core space of the self is basic to the very idea of freedom. Its loss might prove irreversible. It's worth recalling what a philosopher once described as 'the most wretched unfreedom of all':

> To lose the ability even to conceive of what it would be like to have the freedom we lack, and so dismiss even the aspiration to freedom, as something wicked and dangerous.[6]

Was that perhaps what Mark Zuckerberg had in mind when he said that privacy was a thing of the past? Was that what ex-CEO of Google, Eric Schmidt, meant when he wrote that privacy was only for those who had something to hide? They were thinking only about the potential costs for individuals, costs they felt were worth paying (for us, not them). But the real loss will be collective – the loss of freedom as our human horizon. After all, if research under way for machines to 'read' human brains gets implemented on a large scale, we really don't know whether we will have even the physical basis for privacy left.[7] Our brains may always be in reach of scanning machines.

Will this profound shift in power look and feel exactly the same the world over? Not likely. For, as we have emphasised all along, the footprint of the new data colonialism will be overlaid upon the 500-year-old map of historical colonialism, and the extremely unequal economic and political legacy that resulted from it. This inequality of outcomes will play out not only at the individual and group level, but at the level of whole nations. The overwhelming concentration of the dominant digital platforms in the US and China (data colonialism's two main poles) will not shift unless governments and peoples intervene dramatically. But this growing concentration of business and state power through data will make it all the harder for governments outside the US and China to have any effective influence on the regulation of those platforms, including the power to sponsor rival platforms locally. The power play of data colonialism will go on.

To avoid this future, we need local alternatives, some perhaps led by courageous governments, many grounded in specific communities that want a radically different future. That is the only chance of building better platforms whose data collection is driven by social rather than purely commercial goals.

If we are to have the strength to build those alternatives (and Chapter Six gave many starting points), we need to learn from what colonial histories have told us: that a colonial landgrab can change the terms of history comprehensively. Which means it must be *resisted* comprehensively too. This time we will not, unlike the original victims of colonialism, be able to say, 'We didn't know.' We *do* know the history of colonialism, and this book has, we hope, served as a brief reminder of that history and also the history of those who are still resisting it.

The path to resistance will be long and hard, but this book provides a toolkit to get started on that journey. A journey where much is at stake. A journey where what matters most is achieving a more equitable distribution of the world's resources, including the resource we call data. A large collective and imaginative effort will be needed to resist data colonialism's new injustices. But we are hopeful that such an effort can contribute to the longer journey of confronting and ultimately reversing colonialism itself.

NOTES

Introduction – From Landgrab to Data Grab

1 See Neil Parsons, 'The "Victorian Internet" Reaches Halfway to Cairo: Cape Tanganyika Telegraphs 1873–1926', in M. de Bruijn and R. Van Dijck (eds.), *The Social Life of Connectivity in Africa* (Springer, 2012) pp. 95–121; Mosley, Paul, *The Settler Economies: Studies in the Economic History of Kenya and Southern Rhodesia 1900–1963* (Cambridge University Press, 2009) p. 24.

2 Leften S. Stavrianos, *Global Rift: The Third World Comes of Age* (William Morrow and Company, Inc., 1981) pp. 263–4.

3 Platform users as of January 2023: / https://www.statista.com/statistics/272014/global-social-networks-ranked-by-number-of-users/; population figures from https://www.worldometers.info/world-population/.

4 Source: Yahoo, 'GSPC Trending Tickers', *Yahoo Finance* (n.d.) https://uk.finance.yahoo.com/quote/.

5 Bob Pisani, 'Apple versus the world: The iPhone maker is bigger than almost any stock market in the world,' *CNBC* (10 May 2023), https://www.cnbc.com/2023/05/10/apple-vs-the-world-apples-bigger-than-entire-overseas-stock-markets-.html.

6 Anna Gross, Alexandra Heal, Chris Campbell, Dan Clark, Ian Bott and Irene de la Torre Arenas, 'How the US is pushing China out of the internet's plumbing', *Financial Times* (13 June 2013) https://ig.ft.com/subsea-cables/.

7 We developed the concept of data colonialism in a book called *The Costs of Connection* (2019, Stanford University Press) where, if you're interested, you'll find a lot more detail on our ideas. But we are not the first to use the terms 'data grab' or 'data colonialism'. In popular discourse, the first person to use 'data grab', as far as we know, was Dr Neil Bhatia, a doctor in Hampshire, UK, who was referring to a massive drive to centralise data in the National Health Service, which has excited controversy: Laura Clark, ' "Biggest data grab" in NHS history stuffs GP records in a central store for "research" – and the time to opt out is now', *The Register* (13 May 2021), https://www.theregister.com/2021/05/13/nhs_data_grab/; 'The Guardian view on medical records: NHS data grab needs explaining', *Guardian* (30 May 2021), https://www.theguardian.com/commentisfree/2021/may/

30/the-guardian-view-on-medical-records-nhs-data-grab-needs-explaining. For a pioneering academic use of the term, see Alistair Fraser, 'Land Grab, Data Grab: Precision Agriculture and its New Horizons', *Journal of Peasant Studies* 46/5 (2019) pp. 893–912. For anticipations or parallels to our concept of 'data colonialism', see Paul Dourish and Scott D. Mainwaring, 'Ubicomp's Colonial Impulse', *UbiComp '12 Proceedings of the 2012 ACM Conference on Ubiquitous Computing*, (2012) pp. 133–42; Jim Thatcher et al., 'Data colonialism through accumulation by dispossession', *Environment and Planning D: Society and Space*, 34/6 (2016) pp. 990–1006; Renata Avila, 'Digital Sovereignty or Digital Colonialism?', *SUR*, 27 (2018); Antonio Casilli, 'Digital Labor Studies Go Global: Towards a Digital Decolonial Turn,' *International Journal of Communication*, 11 (2017) pp. 3934–54; Julie Cohen, 'The Biopolitical Public Domain' (2017), available from https://papers.ssrn.com/sol3/papers.cfm?abstract_id=2666570, pp. 10–14; Engin Isin and Evelyn Ruppert, 'Data's empire: postcolonial digital politics' in Didier Bigo et al. (eds), *Data Politics: Worlds, Subjects, Rights* (Routledge, 2019) pp. 207–28.

8 Stavrianos, *Global Rift*, p. 154.
9 Philip T. Hoffman, *Why Did Europe Conquer the World?* (Princeton University Press, 2015) p. 2. According to the author: 'Areas under European control here include Europe itself, former colonies in the Americas, and the Russian Empire, but not the non-European parts of the Ottoman Empire.'
10 Americas: David Michael Smith, 'Counting the Dead: Estimating the Loss of Life in the Indigenous Holocaust, 1492–Present', (Southeastern Oklahoma State University, 2018) p. 13. India: Dylan Sullivan and Jason Hickel, 'How British Colonialism Killed 100 Million Indians in 40 Years', Al-Jazeera (2 December 2022). https://www.aljazeera.com/opinions/2022/12/2/how-british-colonial-policy-killed-100-million-indians. Slave trade : Stavrianos, *Global Rift*, p. 109. Algeria : Paul Balta, 'Ombres et lumières de la révolution algérienne', *Le Monde Diplomatique* (1 November 1982), https ://www.monde-diplomatique.fr/1982/11/BALTA/37021. Indonesia: Emmanuel Kreike, 'Genocide in the Kampongs? Dutch nineteenth century colonial warfare in Aceh, Sumatra', *Journal of Genocide Research*, 14:3–4 (2012) pp. 297–315.
11 Stavrianos, *Global Rift*, p. 156.
12 Klas Rönnbäck, 'Sweet Business: Quantifying the Value Added in the British Colonial Sugar Trade in the 18th Century', *Revista de Historia Económica*, 32:2 (2014), pp. 223–45.
13 Sven Beckert, *Empire of Cotton* (Viking, 2015) pp. 328–33.
14 Stavrianos, *Global Rift*, p. 86.
15 Alex Hern, ' "What should the limits be?" The father of ChatGPT on whether AI will save humanity – or destroy it', *Guardian* (7 June 2023), https://www.theguardian.com/technology/2023/jun/07/what-should-the-limits-be-the-father-of-chatgpt-on-whether-ai-will-save-humanity-or-destroy-it.
16 Elizabeth Yale, 'The deep-seated racism of science', *Quartz* (11 March 2016), https://qz.com/637284/the-deep-rooted-racism-of-science.
17 Srinivasa Rao and John Lourdusamy, 'Colonialism and the Development of Electricity: The Case of Madras Presidency, 1900–47', *Science, Technology & Society* 15:1 (2010) pp. 27–54.
18 Achille Mbembe, *Out of the Dark Night* (Columbia University Press, 2021) p 228.

19 'Terms of Service', *Google* (16 April 2007), https://policies.google.com/terms/archive/20051105-20070416?hl=en.

20 *Google Privacy and Terms*, 'Research and development' section (1 July 2023), https://policies.google.com/privacy.

21 We use the terms 'Indigenous' and 'First Nation' interchangeably, in recognition of the ways in which both terms are used by communities and activists. At the same time, we recognise that in some contexts the word 'indigenous' is considered a problematic descriptor.

22 'Spanish Requirement of 1513', *Wikipedia* (last modified 6 January 2023), https://en.wikipedia.org/wiki/Spanish_Requirement_of 1513 [accessed 2 February 2023].

23 Lewis Hanke, *La humanidad es una* (Fondo de Cultura Económica, Mexico City, 1985) p. 70.

24 'Introducing Lasso Blueprint', *Lasso*, (2 February 2022), https://www.lasso-platform.io/post/lasso-blueprint-self-service-audience-builder-for-healthcare, added emphasis.

25 'Google's MusicLM is Astoundingly Good at Making AI-Generated Music, But They're Not Releasing it Due to Copyright Concerns', *AINewsBase* (n.d.), https://ainewsbase.com/google-musiclm-copyright-issues-not-releasing.

26 James Bridle, 'The stupidity of AI', *Guardian* (16 March 2023), https://www.theguardian.com/technology/2023/mar/16/the-stupidity-of-ai-artificial-intelligence-dall-e-chatgpt.

27 Kyle Wiggers, 'Google makes its text-to-music AI public', *TechCrunch* (10 May 2023), https://techcrunch.com/2023/05/10/google-makes-its-text-to-music-ai-public/.

28 Naomi Klein, 'AI machines aren't "hallucinating". But their makers are', *Guardian* (8 May 2023) https://www.theguardian.com/commentisfree/2023/may/08/ai-machines-hallucinating-naomi-klein.

29 Toussaint Nothias, 'Access granted: Facebook's free basics in Africa', *Media, Culture & Society*, 42:3, pp. 329–48. https://doi.org/10.1177/016344371989 0530.

Chapter 1 – A New Colonialism

1 Christiano Lima, 'Google is failing its post-Roe promise to protect abortion privacy', *Washington Post* (24 May 2023) https://www.washingtonpost.com/politics/2023/05/24/google-is-failing-uphold-post-roe-privacy-pledge-democrats-say/.

2 'Colonialism', *Wikipedia*, (last modified 20 January 2023), https://en.wikipedia.org/wiki/Colonialism [accessed 2 February 2023].

3 'Capitalism', *Wikipedia*, (last modified 10 January 2023), https://en.wikipedia.org/wiki/Capitalism [accessed 2 February 2023].

4 For more detail on the facts mentioned in this and the preceding paragraph, see 'Richard Pennant' *Wikipedia*, (last modified 2 January 2023), https://en.wikipedia.org/wiki/Richard_Pennant,_1st_Baron_Penrhyn [accessed 2 February 2023]; on Sir John Pender, see PK Porthcurno, *Changing Places – Porthcurno and the roots of modern communication* (Porthcurno Museum, no date), 8-9; J.M. Blaut, 'Colonialism and the Rise of Capitalism', *Science & Society* 53:3 (1989) pp. 260–96; Robert P. Rogers, 'The Impact of Colonization on Home Country Wealth', *West Palm Beach* 19:4 (2019) pp. 60–76; and for comparison with China which lacked an empire, see Kenneth Pomeranz,

The Great Divergence: China, Europe, and the Making of the Modern World Economy (Princeton University Press, 2000).

5 Karl Marx, *Capital: A Critique of Political Economy Volume 1*, (Penguin Books, 1990).

6 We draw here on Edward E. Baptist, *The Half Has Never Been Told: Slavery and the Making of American Capitalism* (Basic Books, 2016); Sven Beckert, *Empire of Cotton: A New History of Global Capitalism* (Penguin Books, 2015); Sidney W. Mintz, *Sweetness and Power: The Place of Sugar in Modern History* (Penguin Books, 1985); Thomas Piketty and Gabriel Zucman, 'Capital is Back: Wealth-Income Ratios in Rich Countries 1700–2010', *Quarterly Review of Economics* 129:3 (2014) p. 1303; Joshua D. Rothman, *The Ledger and the Chain* (Basic Books, 2021). For a useful summary of these issues, see Matthew Desmond, 'In order to understand the brutality of American capitalism, you have to start with the plantation', *New York Times* (18 August 2019).

7 A. Smith, 'Spatial Division of Labor', in Rob Kitchin and Nigel Thrift (eds), *International Encyclopedia of Human Geography*, (Elsevier Science, 2009), 348–54.

8 For a contemporary perspective on this long debate, see Mirca Madianou, 'Technocolonialism: Digital Innovation and Data Practices in the Humanitarian Response to Refugee Crises', *Social Media & Society* 5/3 (2019): https://doi.org/10.1177/2056305119863146.

9 To acknowledge this, some Marxist theorists have adjusted Marx's term for colonialism's core process ('primitive accumulation') and argued that it continues throughout capitalism, as 'accumulation by dispossession' or expropriation: David Harvey, 'The "New Imperialism": Accumulation by dispossession', *Socialist Register* 40 9 (2004): 63–87; Nancy Fraser, *Cannibal Capitalism* (Verso 2022), chapters 1 and 2. We don't disagree on the underlying facts, but why not draw the obvious conclusion that capitalism always works, in part, through *colonial* means? That said, we do agree with Fraser when she writes that 'expropriation . . . usually seen as the antithesis of capitalism signature process of exploitation . . . is better seen as the latter's enabling condition' (ibid.: 14–15).

10 See for instance the theories of digital capitalism – Dan Schiller, *Digital Depression: Information Technology and Economic Crisis* (University of Illinois Press, 2014); platform capitalism – Nick Srnicek, *Platform Capitalism* (Wiley, 2016); data capitalism – Viktor Mayer-Schonberger and Thomas Ramge, *Reinventing Capitalism in the Age of Big Data* (John Murray, 2018); informational capitalism – Julie Cohen, *Between Truth and Power: The Legal Constructions of Informational Capitalism* (Oxford University Press, 2019); and surveillance capitalism – Shoshana Zuboff, *The Age of Surveillance Capitalism: The Fight for a Human Future at the New Frontier of Power* (Public Affairs, 2019).

11 Zuboff, *The Age of Surveillance Capitalism*.

12 Achille Mbembe, *Critique of Black Reason*, (Duke University Press, 2017): we'll come back to this important point in Chapter 5.

13 We are summarising here rationalisations that took decades and centuries to evolve during the first, historical phase of colonialism. Source: Anthony Pagden, *Lords of All the World: Ideologies of Empire in Spain, Britain, and France c. 1500–1800* (Yale Books, 1998). More on this in Chapter 3.

14 Partha Chatterjee, *The Black Hole of Empire: History of a Global Practice of Power* (Princeton University Press, 2012) pp. 47, 52, 146, 162.

15 Stavrianos, p. 156.

16 Pierre Bourdieu, *Language and Symbolic Power* (Polity, 1990); more recently, Anna Lauren Hoffmann, 'Terms of Inclusion: Data, discourse, violence', *New Media and Society* 23:12 (2021), 3539–56.

17 Keith Breckenridge, *Biometric State: The Global Politics of Identification and Surveillance in South Africa, 1850 to the Present* (Cambridge University Press, 2014).

18 Oscar Gandy Jr, *The Panoptic Sort: A Political Economy of Personal Information* (Oxford University Press, 2021); Wendy Hui Kyong Chun, *Discriminating Data: Correlation, Neighborhoods, and the New Politics of Recognition* (MIT Press, 2021).

19 Virginia Eubanks, *Automating Inequality: How High-Tech Tools Profile, Police, and Punish the Poor* (St. Martin's Press, 2018); Mary Madden et al., 'Privacy, Poverty and Big Data: A Matrix of Vulnerabilities for Poor Americans', *Washington University Law Review* 95:1 (2017) pp. 53–125; Julia Ticona, *Left to Our Own Devices: Coping with Insecure Work in a Digital Age* (Oxford University Press, 2022); Seeta Peña Gangadharan, 'Digital Inclusion and Racial Profiling', *First Monday* 17:5 (2012).

20 Karen Hao, 'The Facebook whistleblower says its algorithms are dangerous. Here's why', *MIT Technology Review*, (5 October 2021) https://www.technologyreview.com/2021/10/05/1036519/facebook-whistleblower-frances-haugen-algorithms/.

21 Dahaba Ali Hussen, ' "Dystopian" surveillance disproportionately targets young, female and minority workers', *Guardian* (26 March 2023), https://www.theguardian.com/global-development/2023/mar/26/dystopian-surveillance-disproportionately-targets-young-female-minority-workers-ippr-report.

22 Sheelah Kolhatkar, 'The Fight to Hold Pornhub Accountable', *New Yorker* (13 June 2020),https://www.newyorker.com/magazine/2022/06/20/the-fight-to-hold-pornhub-accountable.

23 G. Macías, 'The Streets Are Ours: The Violence of Surveillance Against Women in Public Space'. In A. Argüelles (ed.), *Digital Violence in México: The State vs Civil Society*, (Comun.al, 2022) 63–88.

24 Digital Health and Rights Project Consortium, 'Digital health and human rights of young adults in Ghana, Kenya, and Vietnam: Final project report', *Geneva Graduate Institute* (2022), https://www.graduateinstitute.ch/library/publications-institute/digital-health-and-human-rights-young-adults-ghana-kenya-and-vietnam.

25 Gina Neff, 'The Internet is at Risk of Driving Women Away', *Wired* (19 December 2022), https://www.wired.com/story/online-harassment-women-internet/?s=03.

26 Dhanaraj Thakur and DeVan Hankerson Madrigal, 'An Unrepresentative Democracy: How Disinformation and Online Abuse Hinder Women of Color Political Candidates in the United States', *Center for Democracy and Technology*, (27 October 2022), https://cdt.org/insights/an-unrepresentative-democracy-how-disinformation-and-online-abuse-hinder-women-of-color-political-candidates-in-the-united-states/.

27 Ayana Archie, 'Two women who allege they were stalked and harassed using AirTags are suing Apple', *NPR*, (7 December 2022), https://www.npr.org/2022/12/07/1141176120/apple-airtag-harassment-stalker-lawsuit.

28 Anna Lauren Hoffmann, 'Terms of Inclusion: Data, discourse, violence', *New Media & Society* 23:2 (2020): pp. 3539–56.

29 Catherine D'Ignazio and Lauren F. Klein, *Data Feminism* (The MIT Press, 2020) pp. 8–9.
30 Heather Vogell, 'Rent Going Up? One Company's Algorithm Could Be Why', *ProPublica* (15 October 2022), https://www.propublica.org/article/yield star-rent-increase-realpage-rent.
31 Ziad Obermeyer et al., 'Dissecting racial bias in an algorithm used to manage the health of populations', *Science* 366:6464 (2019), 447–53.
32 Efrén Cruz Cortés et al., 'Locality of Technical Objects and the Role of Structural Interventions of Systemic Change', *2022 ACM Conference on Fairness, Accountability, and Transparency (FAccT '22)* (2022) pp. 2327–41; Dan McQuillan, *Resisting AI: An Anti-fascist Approach to Artificial Intelligence* (Bristol University Press, 2022).
33 Steven J. Harris, 'Long-Distance Corporations, Big Sciences, and the Geography of Knowledge', in Sandra Harding, e(ds), *The Postcolonial Science and Technology Studies Reader* (Duke University Press, 2011) pp. 61–83.
34 See Steven J. Harris, 'Long-Distance Corporations, Big Sciences, and the Geography of Knowledge', pp. 61–83; as well as Stefanie Gänger, 'World Trade in Medicinal Plants from Spanish America', *Medical History* 59:1 (2015) pp. 44–62.
35 See 'List of pre-Columbian inventions and innovations of indigenous Americans', *Wikipedia* (last modified 8 December 2022) https://en.wikipedia.org/w/index.php?title=List_of_pre-Columbian_inventions_and_innovations_of_indigenous_Americans [accessed 2 February 2023]; and Patrick J. Kiger, '10 Native American Inventions Commonly Used Today', (14 November 2019) https://www.history.com/news/native-american-inventions.
36 Stavrianos, p. 559.
37 See Ahmed H. Sa'di, 'Colonialism and Surveillance,' in Kirstie Ball et al., (eds), *Routledge Handbook of Surveillance Studies* (Routledge, 2012), 151–8; Keith Breckenridge, *Biometric State*.
38 Alan Blackwell, 'Ethnographic artificial inteliigence', *Interdisciplinary Science Reviews* 46/1-2 (2021): 198-211, at 204.
39 Therese Poletti and Jeremy C. Owens, '$1.4 trillion? Big Tech's pandemic year produces mind-boggling financial results', *Market Watch* (7 February 2022), https://www.marketwatch.com/story/1-4-trillion-big-techs-pandemic-year-produces-mind-boggling-financial-results-11644096594.
40 Bob Violino, 'Data privacy rules are sweeping across the globe, and getting stricter', *CNBC* (22 December 2022), https://www.cnbc.com/2022/12/22/data-privacy-rules-are-sweeping-across-the-globe-and-getting-stricter.html.
41 Patrick McGee, 'Meta and Alphabet lose dominance over US digital ads market', *Ars Technica* (23 December 2022), https://arstechnica.com/tech-policy/2022/12/meta-and-alphabet-lose-dominance-over-us-digital-ads-market/.
42 Christina Criddle, 'Meta agrees $775 mn to settle Cambridge Analytica Case', *Financial Times* (23 December 2022), https://www.ft.com/content/52cebe72-1894-4679-ac51-47d48171f6f6; Hannah Murphy and Javier Espinosa 'Meta fined almost €400 mn over EU privacy violations', (4 January 2023), https://www.ft.com/content/5f8240af-865a-4cf6-8a5a-4d6cf1d5fcb1.
43 Joe Panettieri, 'Big Tech Antitrust Investigations: Amazon, Apple, Google, Meta/Facebook and Microsoft Updates', *Channel E2E*, (5 December 2022) https://www.channele2e.com/business/compliance/big-tech-antitrust-regulatory-breakup-updates/.

44 Tabby Kinder, 'Silicon Valley start-ups race for debt deals in funding crunch', *Financial Times* (20 December 2022), https://www.ft.com/content/d12a4baa-53c6-4256-b125-d25622f9555f.

45 Robert Armstrong, 'Were we wrong about big tech?' *Financial Times* (7 November 2022) https://www.ft.com/content/e9098953-e031-4bc5-9bc4-d5341db8f2f9.

46 Keerthi Vedantam, 'Tech Layoffs: US Companies That Cut Jobs In 2022 and 2023', *Tech Crunch* (31 March 2023), https://news.crunchbase.com/start-ups/tech-layoffs.

47 Tech jobs in the US overall *increased* by 12% in the US during 2022: see John Thornhill, 'In (partial) defence of Silicon Valley', *Financial Times* (5 January 2023), https://www.ft.com/content/ae6575ea-2af4-4557-bce2-f924dca7d676.

48 Jack Kelly, 'Layoff Contagion Spreads From The Tech Sector To Everywhere Else', *Forbes* (8 December 2022), https://www.forbes.com/sites/jackkelly/2022/12/08/layoff-contagion-spreads-from-the-tech-sector-to-everywhere-else/?sh=384590382fe4.

49 Manuel Ayala and Carlos A. Moreno, 'Amazon was supposed to transform a Tijuana slum. It failed to deliver', *Rest of The World* (12 December 2022), https://restofworld.org/2022/amazon-tijuana-slum-promises/.

50 Naomi Klein, *The Shock Doctrine* (Penguin Books, 2008).

51 Miriyam Aouragh, Seda Gürses, Helen Pritchard, and Femke Snelting, 'The extractive infrastructures of contact tracing apps', *Journal of Environmental Media* 1 (Supplement) (2020), https://westminsterresearch.westminster.ac.uk/download/f20a6a448558aee12e9f10684c6f3d84a424c31892c4df-1614247b5a4b101a40/117883/s10.pdf.

52 'Under Surveillance: (Mis)use of Technologies in Emergency Responses Global lessons from the Covid-19 pandemic', *Privacy International* (14 December 2022), https://privacyinternational.org/report/5003/under-surveillance-misuse-technologies-emergency-responses-global-lessons-covid-19.

53 Ian Bogost, 'The Age of Social Media is Ending', *The Atlantic* (10 November 2022), https://www.theatlantic.com/technology/archive/2022/11/twitter-facebook-social-media-decline/672074/; Morgan Meaker, 'The Slow Death of Surveillance Capitalism Has Begun', *Wired* (5 January 2023), https://www.wired.co.uk/article/meta-surveillance-capitalism.

Chapter 2 – Data Territories

1 Giancarlo Spagnolo, Otto Toivanen, Nicolas Schutz, Michelle Sovinsky, Thomas Rønde, Monika Schnitzer, Paul Heidhues, Cristina Caffarra, Tomaso Duso, Christos Genakos, Gregory Crawford, Chongwoo Choe, Zhijun Chen, Tommaso Valletti, Martin Peitz, Thibaud Vergé and Marc Bourreau, 'Google/Fitbit will monetise health data and harm consumers', *VoxEU* (30 September 2020), https://cepr.org/voxeu/columns/googlefitbit-will-monetise-health-data-and-harm-consumers.

2 Jason Peres da Silva, 'Privacy Data Ethics of Wearable Digital Health Technology', *Warren Alpert Medical School Center for Digital Health* (4 May 2023), https://digitalhealth.med.brown.edu/news/2023-05-04/ethics-wearables; Alexia Dini Kounoudes, Georgia M. Kapitsaki & Ioannis Katakis, 'Enhancing user awareness on inferences obtained from fitness trackers data' *User Modeling and User-Adapted Interaction* (17 January 2023), https://doi.

org/10.1007/s11257-022-09353-8; David Cox, 'The rise of employee health tracking,' *BBC Worklife* (11 November 2020), https://www.bbc.com/worklife/article/20201110-the-rise-of-employee-health-tracking.

3 https://www.ftc.gov/enforcement/refunds/equifax-data-breach-settlement.

4 Sources: https://www.britannica.com/list/8-of-the-largest-empires-in-history.

5 Eric Jones, *The European Miracle: Environments, Economies, and Geopolitics in the History of Europe and Asia* (Cambridge University Press, 1987), discussed in Dimitrios Theodoridis, 'Colonialism and Trade: Ecological Foundations of British Trade in the Nineteenth Century', *Journal of Interdisciplinary History* 53:1 (2022) pp. 1–23; see also Kenneth Pomeranz, *The Great Divergence* (Princeton University Press, 2001).

6 An anticipation of this idea was critiques in the 2000s of the 'enclosure' of the internet commons by commercial computing infrastructures: James Boyle, 'The Second Enclosure Movement and the Construction of the Public Domain', *Law and Contemporary Problems* 66 (2003) pp. 147–78; Mark Andrejevic, 'Surveillance in the Digital Enclosure', *The Communication Review* 19/4 (2007) pp. 295–317.'

7 Philip Agre, 'Surveillance and Capture: Two Models of Privacy', *The Information Society* 10:2 (1994), 101–27.

8 Lawrence Lessig (1999) *Code and Other Laws of Cyberspace* (Basic Books, 1999); Joel Reidenberg, 'Lex Informatica: The Formulation of Information Policy Rules through Technology', *Texas Law Review* 76 (1997–8), 553–93 (available at http://ir.lawnet.fordham.edu/faculty_scholarship/42); Wendy Chun, *Programmed Visions* (MIT Press, 2011).

9 Paul Schwartz 'Internet Privacy and the State', *Connecticut Law Review* 32 (2000) pp. 815–59; Julie E. Cohen 'Examined Lives: Informational Privacy and the Subject as Object', *Stanford Law Review* 52 (2000) pp. 1373–438, available from https://scholarship.law.georgetown.edu/facpub/810/; Oscar Gandy, *Panoptic Sort*.

10 Hal R. Varian 'Beyond Big Data', *Paper presented at the NABE Annual Meeting, San Francisco* (2013), http://edshare.soton.ac.uk/15212/7/BeyondBigDataPaperFINAL.pdf.

11 Hal R. Varian, 'Beyond Big Data', p. 2.

12 Ibid., p. 8.

13 Srnicek, *Platform Capitalism*.

14 Alice Hearing, 'Meta's Horizon World's quality is so poor not even employees are using it, says VP in leaked memo', *Fortune* (7 October 2022), https://fortune.com/2022/10/07/metas-horizon-world-quality-poor-not-even-employees-using-says-metaverse-vp-leaked-memo/.

15 Nicolaj Siggelkow and Christian Terwiesch, 'The Age of Continuous Connection', *Harvard Business Review* (2019) https://hbr.org/2019/05/the-age-of-continuous-connection; Mercedes Bunz and Graham Meikle, *Internet of Things* (Wiley, 2017).

16 ACLU, 'Submission to Federal Trade Commission on Commercial Surveillance', *Federal Trade Commission*, (21 November 2022), https://www.regulations.gov/comment/FTC-2022-0053-1185.

17 Tanya Basu, 'Amazon has a new plan for its robot: To guard your life', *MIT Tech Review* (28 September 2022), https://www.technologyreview.com/2022/09/28/1060418/amazon-wants-astro-to-guard-your-life/.

18 'Amazon is buying iRobot for $1.7 billion', *TechCrunch* (5 August 2022), https://www.npr.org/2022/08/05/1115946395/amazon-buying-roomba-irobot.

19 Yong Jin Park, *The Future of Digital Surveillance* (Michigan University Press, 2021).

20 Julie Cohen, *Between Truth and Power*, p. 38.

21 Eileen Guo, 'A Roomba recorded a woman on the toilet. How did screenshots end up on Facebook?', *MIT Tech Review* (19 December 2022) https://www.technologyreview.com/2022/12/19/1065306/roomba-irobot-robot-vacuums-artificial-intelligence-training-data-privacy/.

22 Tate Ryan Mosley, 'How to hack a smart fridge', *MIT Tech Review* (8 May 2023) https://www.technologyreview.com/2023/05/08/1072708/hack-smart-fridge-digital-forensics/.

23 Thomas Davenport, *Big Data @ Work* (Harvard Business Review Press, 2014) p. 11; Thomas M. Siebel, *Digital Transformation* (Rosetta Books, 2019) p. 18.

24 Hal R. Varian, 'Artificial, Intelligence, Economics and Industrial Organization', in Ajay Agrawal, Joshua Gans and Avi Goldfarb (eds), *The Economics of Artificial Intelligence* (Chicago University Press, 2019) p. 328.

25 Simone Browne, *Dark Matters* (Duke University Press, 2015); Edward E. Baptist, *The Half Has Never Been Told*.

26 Shoshana Zuboff, *The Age of Surveillance Capitalism*, pp. 63–97; Karen Yeung, '"Hypernudge": Big Data as a Mode of Regulation by Design', *Information Communication & Society* 20:1 (2017) pp. 118–36.

27 Meng Liang, 'The end of social media? How data attraction model in the algorithmic media reshapes the attention economy', *Media Culture & Society* 44:6 (2022) pp. 1110–31.

28 Hannah Murphy and Cristina Criddle, 'Meta's AI-driven advertising system splits marketers', *Financial Times* (27 February 2023), https://www.ft.com/content/fc95a0f7-5e4e-4616-9b17-7b72daee6c60.

29 Mark Andrejevic, *Infoglut* (Routledge, 2013).

30 Anne Helmond, 'The Platformization of the Web: Making Web Data Platform Ready', *Social Media & Society*, 1/2 (2015) pp. 1–11.

31 Tobias Blanke and Jennifer Pybus, 'The Material Conditions of Platforms: Monopolization through Decentralization', *Social Media & Society* (2020), DOI: 10.1177/2056305120971632.

32 Quoted, Joseph Turow, *The Voice Catchers* (Yale University Press, 2021) p. 27.

33 Ryan Mac, Charlie Warzel and Alex Kantrowitz, 'Growth at Any Costs: Top Facebook Executive Defended Data Collection in 2016 Memo – and Warned that Facebook Could Get People Killed', *BuzzFeed* (29 March 2018), https://www.buzzfeednews.com/article/ryanmac/growth-at-any-cost-top-facebook-executive-defended-data, added emphasis.

34 Neil C. Thompson et al., 'The Computational Limits of Deep Learning', *Machine Learning* (2022) pp. 1–23.

35 Richard Maxwell and Toby Miller, *Greening the Media* (Oxford University Press, 2012) p. 95; Julia Velkova, 'Data That Warms: waste heat, infrastructural convergence and the computation traffic commodity, *Big Data & Society* 3:2 (2016) p. 4; Lauren Bridges, 'Toxic Clouds and Dirty Data', (10 October 2022) *Kleinman Center for Energy Policy*, https://kleinmanenergy.upenn.edu/news-insights/toxic-clouds-and-dirty-data/.

36 George Hammond and Stephen Morris, 'West London faces new homes ban as electricity grid hits capacity', *Financial Times* (28 July 2022) https://www.ft.com/content/519f701f-6a05-4cf4-bc46-22cf10c7c2c0; Alexandra Heal and Anna Gross, 'Thames Water reviews data centres' water use as London hosepipe ban looms', *Financial Times* (23 August 2022) https://www.ft.com/content/8d8bf26f-5df2-4ff6-91d0-369500ed1a9c.

37 TeleGeography, *The State of the Network 2023 edition*, 15, available from https://www2.telegeography.com/download-state-of-the-network.

38 Benjamin A. Jones, Andrew L. Goodkind and Robert P. Berrens, 'Economic estimation of Bitcoin mining's climate damages demonstrates closer resemblance to digital crude than digital gold', *Scientific Reports* 12, (2022) 14512.

39 Nathan Ensmenger, 'The Environmental History of Computing', *Technology and Culture* 59:4 (2018), S7–S33.

40 Richard Maxwell and Toby Miller, *Greening the Media*, p. 93.

41 Kate Crawford and Vladan Joler, 'Anatomy of an AI System', (2018) https://anatomyof.ai/; Kate Crawford, *Atlas of AI* (Yale University Press, 2021).

42 Sebastian Lehuedé, *Mobilising Water: Elemental Resistance in the Technocene*. Presentation, IAMCR (July 2023), Lyon, France.

43 Vanessa Forti et al., *The Global E-waste Monitor 2020: Quantities, flows and the circular economy potential*. United Nations Institute for Training and Research, (2020) https://ewastemonitor.info/gem-2020/.

44 John Gapper, 'Material World – the six commodities that shape our lives' [review of Ed Conway, *Material World*], *Financial Times* (15 June 2023), https://www.ft.com/content/4131123b-8403-4151-8c3e-5d6fadc1df45. And see Further Reading.

45 Klein, *This Changes Everything* (Penguin Books, 2015) pp. 172–3.

46 For background, Joseph Turow, *The Daily You: How the New Advertising Industry Is Defining your Identity and Your Worth* (Yale University Press, 2011); Tim Hwang, *Subprime Attention Crisis* (Macmillan Publishers, 2019). For Apple, see Patrick McGee, 'Apple Plans to double its digital advertising business workforce', *Financial Times* (5 September 2022) https://www.ft.com/content/db21685b-d4dd-421d-95ac-980e9d40c05c.

47 Morgan Meaker, 'The Slow Death of Surveillance Capitalism Has Begun', *Wired* (5 January 2023) https://www.wired.co.uk/article/meta-surveillance-capitalism.

48 Marion Fourcade and Kieran Healy, 'Classification Situations: life-chances in the neoliberal era', *Accounting, Organizations and Society* 38:8 (2013) pp. 559–72.

49 Bayer, 'Technology Stewardship Agreement: Frequently Asked Questions', *Bayer*, (n.d.) https://traits.bayer.com/stewardship/Documents/tsa-faqs-stewardship.pdf.

50 Tencent, 'Bridging Gaps in Healthcare Industry with Technology', *Tencent*, (11 August 2019) https://www.tencent.com/en-us/articles/2200933.html; Xiaowei Weng, 'Behind China's Pork Miracle,' *Guardian* (8 October 2020), https://www.theguardian.com/environment/2020/oct/08/behind-chinas-pork-miracle-how-technology-is-transforming-rural-hog-farming. For discussion, see Anita Gurumurthy and Nandini Chami, 'The Intelligent Corporation,' *TNI Longreads* (16 January 2020), https://longreads.tni.org/stateofpower/the-intelligent-corporation-data-and-the-digital-economy.

51 Gurumurthy and Chami, *The Intelligent Corporation* 12, on global dairy production and Alibaba.

52 Cited Kelly Bronson, 'The Immaculate Conception of Data: Agribusiness, activists, and their shared politics of the future' (McGill-Queen's University Press, 2022) p. 35.

53 Kelly Bronson and Irena Knezevic, 'Big Data in Food and Agriculture', *Big Data & Society* 3:1 (2016) pp. 1–5; Christopher Miles, 'The Combine will Tell the Truth', *Big Data & Society* 6:1 (2019) pp. 1–12; Alistair Fraser 'Land Grab/Data Grab', *Journal of Peasant Studies* 46:5 (2019) pp. 893–912.

54 Jon Brodkin, 'Colorado governor signs tractor right-to-repair law opposed by John Deere', *Ars Technica* (26 April 2023), https://www.apple.com/uk/newsroom/2022/12/apple-launches-self-service-repair-in-europe/. For Apple's recent reversal of its block on the independent right to repair, see https://www.apple.com/uk/newsroom/2022/12/apple-launches-self-service-repair-in-europe/.

55 Kelly Bronson, *Immaculate Conception*, pp. 35–7, 120.

56 'Irish Inventors Driving the High-tech Farm of the Future', *Independent* (24 May2015),http://www.independent.ie/business/technology/irish-inventors-driving-thehightech-farm-of-the-future-31248410.html; Science Gallery Dublin, 'Grassometer', *Trinity College Dublin* (n.d.), https://dublin.sciencegallery.com/field-test-exhibits/grassometer.

57 Kristen E. DiCerbo and John T. Behrens, *Impacts of the Digital Ocean on Education* (February 2014), https://www.pearson.com/content/dam/one-dot-com/one-dot-com/global/Files/about-pearson/innovation/open-ideas/DigitalOcean.pdf.

58 Louise Hooper, Sonia Livingstone, Krakae Pothong, 'Problems with Data Governance in UK schools: the cases of Google Classroom and ClassDojo', *Digital Futures Commission and 5Rights Foundation* (August 2022), https://digitalfuturescommission.org.uk/wp-content/uploads/2022/08/Problems-with-data-governance-in-UK-schools.pdf, p. 10. See generally Ben Williamson, *Big Data in Education* (Sage Publications, 2017).

59 Bethan Staton, 'Education companies' shares fall sharply after warning over ChatGPT', *Financial Times* (2 May 2023) https://www.ft.com/content/0db12614-324c-483c-b31c-2255e8562910.

60 Notícias Corporativas, 'Número de edtechs cresce 26% no Brasil durante a pandemia', *Mundo do Marketing* (n.d.), https://www.mundodomarketing.com.br/noticias-corporativas/conteudo/271870/numero-de-edtechs-cresce-26-no-brasil-durante-a-pandemia; David Pilling, 'Lagos turns to edtech partner to raise school standards', *Financial Times* (15 February 2022), https://www.ft.com/content/e294aa8e-710c-4524-9cc3-0060703f2252.

61 Human Rights Watch, 'How Dare they Peep into my Private Life?', *Human Rights Watch* (2022), https://www.hrw.org/report/2022/05/25/how-dare-they-peep-my-private-life/childrens-rights-violations-governments, pp. 103–24.

62 Selena Nemorin et al., 'AI Hyped: A Horizon scan of discourse on artificial intelligence in education (AIED) and development', *Learning Media and Technology* 48:1 (2023) p. 5, citing 'Audrey Azoulay: Making the Most of Artificial Intelligence', *UNESCO Courier* 3 (2018) https://en.unesco.org/courier/2018-3/audrey-azoulay-making-most-artificial-intelligence; EU High-Level Expert Group on AI (AIHLEG), 'Ethics Guidelines for Trustworthy AI,' *European Commission* (8 April 2019), https://digital-strategy.ec.europa.eu/en/library/ethics-guidelines-trustworthy-ai.

63 Quoted Nemorin et al., 'AI Hyped', p. 7.

64 Hooper et al., 'Problems with Data Governance'.
65 Hooper et al., 'Problems with Data Governance', pp. 55–6.
66 Human Rights Watch, *How Dare they Peep?*
67 Stephen Hutt et al., 'Automated Gaze-Based Mind Wandering Detection during computerized learning in classrooms', *User Modeling and User-Adapted Interaction* 29 (2019) pp. 821–67, 822, 823.
68 Faye J. Jones, 'SLANT: A New Behavioural Management System', *National Association of Special Education Teachers*, (n.d.), https://www.naset.org/publications/classroom-management-series/slant-a-new-behavior-management-system.
69 Chloe Cornish, Jyotsna Singh and Mercedes Ruehl, 'How a teaching app feted by Silicon Valley was left chasing the Indian dream', *Financial Times* (3 October 2022) https://www.ft.com/content/8d998d44-3937-4931-8ee1-f963f6a8b253.
70 Michael Veale, 'Schools Must Resist Big EdTech – but it won't be easy', in Sonia Livingstone and Krakae Pothong (eds), *Education Data Futures* (Digital Futures Commission, 2022) pp. 66–79.
71 Quotes from Center for Digital Democracy and 19 others, Submission to Federal Trade Commission's Advanced notice of Proposed Rulemaking on Commercial Surveillance (21 November 2022), available from https://www.regulations.gov/comment/FTC-2022-0053-1144.
72 Ben Williamson 'PIS for machine learners', blog (25 November 2021), https://codeactsineducation.wordpress.com/2021/11/25/pisa-for-machine-learners/; Velislava Hillman, 'Algorithmic Systems Claim Education: And (Re)Production of Societies', in P. Jandrič, A. MacKenzie and J. Knox (eds), *Constructing Postdigital Research: Method and Emancipation* (Springer Nature, forthcoming 2023).
73 Julia Powles and Hal Hodson, 'Google DeepMind and healthcare in an age of algorithms', *Health Technology Health Technology* 7 (2017) pp. 351–67.
74 Emma Roth, 'Cerebral admits to sharing patient data with Meta, TikTok, and Google', *The Verge* (11 March 2023), https://codeactsineducation.wordpress.com/2021/11/25/pisa-for-machine-learners/.
75 Daniel Sulmasy, 'Naked bodies, naked genomes: the special (but not exceptional) nature of genomic Information', *Genetics in Medicine* 17 (2015) pp. 331–6.
76 '23 and me', *23andMe* (n.d.), https://www.23andme.com/; See Jamie Ducharme, 'A Major Drug Company Now Has Access to 23andMe's Genetic Data: Should You be Concerned?' *Time* (26 July 2018), https://time.com/5349896/23andme-glaxo-smith-kline/.
77 Lex column, 'Mental health apps: the AI therapist cannot see you now', *Financial Times* (4 June 2023) https://www.ft.com/content/e0730064-7b24-4556-9322-45a806c4c5f7.
78 John Glaser et al., 'How to Use Digital Health Data to Improve Outcomes', *Harvard Business Review* (12 September 2022), https://hbr.org/2022/09/how-to-use-digital-health-data-to-improve-outcomes; Alex Pentland, Alexander Lipton and Thomas Hardjono, *Building the New Economy: Data as Capital* (MIT Books, 2022), Chapter 7.
79 Ronald J. Deibert, 'The Autocrat in your iPhone: How Mercenary Spyware Threatens Democracy', *Foreign Affairs* (January/February 2023), https://www.foreignaffairs.com/world/autocrat-in-your-iphone-mercenary-spyware-ronald-deibert.

80 Kevin B. Johnson et al., 'Precision Medicine, AI and the Future of Personal-ized Health Care', *Clinical and Translational Science* 14/1 (2021) pp. 86–93, https://doi.org/10.1111/cts.12884.

81 CIO Tech Team, 'HK Sees Internet of Things as the Core to the Develop-ment of Digital Economy', *CIO Tech Asia* (19 January 2022), https://ciotechasia.com/hk-sees-internet-of-things-as-the-core-to-the-development-of-digital-economy/.

82 Ivan Mehta, 'India Pips North America to Become the Biggest Smartwatch Market', *TechCrunch* (30 November 2022).

83 Sarah E. Needleman and Rob Copeland, 'Google Counts on Fitbit to Make Imprint in Health Market', *The Wall Street Journal* (6 November 2019) https://www.wsj.com/articles/google-counts-on-fitbit-to-make-imprint-in-health-market-11573052061.

84 'How Big Tech Captured Our Public Health System', interview by Arun Kundnani with Seda Gürses (June 2021) https://www.tudelft.nl/2022/tbm/how-big-tech-captured-our-public-health-system.

85 Tencent, 'Bridging Healthcare Gaps'.

86 Aynne Kokas, *Trafficking Data* (Oxford University Press, 2022) p. 158; R. Liao, 'TikTok parent ByteDance just bought a hospital group in China', *TechCrunch* (9 August 2022), https://tcrn.ch/3bFCIOV.

87 *Lancet*, 'The Lancet and the Financial Times Commission on Governing Health Futures 2030: growing up in a digital world', *The Lancet* (24 October 2021), https://www.thelancet.com/commissions/governing-health-futures-2030, p. 2.

88 https://www.hippoai.org/. See also this interview with Hippo AI's founder, Bart de Witte in *Die Ärtztekammer Steiermark*, (7 August 2023), https://www.aekstmk.or.at/507?articleId=12625.

89 Karl Marx, *Capital: A Critique of Political Economy Volume 1* (Penguin Books, 1976) p. 449.

90 Baptist, *The Half Has Never Been Told*; Browne, *Dark Matters*.

91 Ifeoma Ajunwa, *The Quantified Worker: Law and Technology in the Modern Workplace* (Cambridge University Press, 2023).

92 '6 in 10 employers require monitoring software for remote workers', *Digital* (4 October 2021), https://digital.com/6-in-10-employers-require-monitoring-software-for-remote-workers/; Andrew Fennell, 'How much does your boss know about you?', *Standout CV* (2023), https://standout-cv.com/employee-monitoring-study.

93 Julia Ticona, *Left to Our Own Devices: Coping with Insecure Work in a Digital Age* (Oxford University Press, 2022).

94 David Weil, *The Fissured Workplace* (Harvard University Press, 2014).

95 Anja Kanngieser, 'Tracking and Tracing: geographies of logistical governance and labouring bodies', *Environment and Planning D: Society and Space* 31/4 (2012) pp. 594–610, 602, citing GMB Union 2006 report on 'Health and Safety at Work: R F voice pick survey'.

96 Quoted, Heather Stewart, ' "We're Not Going Away": UK strike trio bullish over battle for Amazon union', *Guardian* (12 June 2023), https://www.theguardian.com/technology/2023/jun/11/amazon-coventry-strike-trio-bullish-union.

97 International Brotherhood of Teamsters, 'Comment submitted by Inter-national Brotherhood of Teamsters', Federal Trade Commission (21 November 2022), https://www.regulations.gov/comment/FTC-2022-0053-1223.

98 Karen E.C. Levy, 'The Contexts of Control: Information, power, and truck-driving work', *The Information Society* 31:2 (2015) pp. 160–74, 167.
99 Quoted, Levy, 'Contexts of Control', p. 166.
100 Mary Madden et al., 'Privacy, Poverty and Big Data: A Matrix of Vulnerabilities for Poor Americans', *Washington University Law Review* 95:1 (2017) pp. 53–125.
101 TUC, 'Gig economy workforce in England and Wales has almost tripled in last five years – new TUC research', (5 November 2021), https://www.tuc.org.uk/news/gig-economy-workforce-england-and-wales-has-almost-tripled-last-five-years-new-tuc-research; on US freelancers, see https://www.statista.com/statistics/921593/gig-economy-number-of-freelancers-us/.
102 Gurumurthy and Chami, 'The Intelligent Corporation'.
103 Lilly Irani, Mikhail Hussain, Peter Zschiesche et al., 'Transportation for Smart and Equitable Cities: Integrating Taxis and Mass Transit for Access, Emissions Reduction, and Planning', (UC San Diego Design Lab, 2021). https://escholarship.org/uc/item/06w4d22x.
104 Mary Gray and Siddharth Suri, *Ghostwork* (Houghton Mifflin Harcourt, 2019).
105 Ryan Calo and Alex Rosenblat, 'The Taking Economy', *Columbia Law Review* 117:6 (2014) pp. 4–5.
106 Gray and Suri, *Ghostwork*, p. 34.
107 'Uber, Lyft Sink after Biden Administration proposes new gig work rule', *Financial Times,* (11 October 2022), https://www.ft.com/content/9af840e4-7494-4d8d-ab25-a3d1de8ecf20.
108 Kathleen Griesbach et al., 'Algorithmic Control in Platform Food Delivery Work', *Socius: Sociological Research for a Dynamic World* 5 (2019) pp. 1–15.
109 James Muldoon and Paul Raekstad, 'Algorithmic Domination in the Gig Economy', *European Journal of Political Theory* (2022), https://doi.org/10.1177/14748851221082078.
110 Kate Vredebergh, 'Freedom at Work', *Canadian Journal of Philosophy* 51:1 (2022) pp. 78–92; McQuillan, *Resisting AI*, pp. 53-54; Karen Hao and Andrea Paola Hernández, 'How the AI Industry Profits from Catastrophe', *MIT Tech Review* (20 April 2022), https://www.technologyreview.com/2022/04/20/1050392/ai-industry-appen-scale-data-labels/.
111 Monica Anderson et al., 'The State of Gig Work in 2021', *Pew Research Center* (8 December 2021), https://www.pewresearch.org/internet/2021/12/08/the-state-of-gig-work-in-2021/.
112 Nicole Starosielski, *The Undersea Network* (Duke University Press, 2015).
113 Sofia Scasserra and Carolina M. Elebi, 'Digital Colonialism: Analysis of Europe's Trade Agenda', *TNT* (6 October 2021), https://www.tni.org/en/publication/digital-colonialism.

Chapter 3 – Data's New Civilising Mission

1 Anna Lena Hunkenschroer and Alexander Kriebitz, 'Is AI recruiting (un)ethical? A human rights perspective on the use of AI for hiring', *AI Ethics* 3, pp. 199–213 (25 July 2023), https://doi.org/10.1007/s43681-022-00166-4; Jonathan Barrett and Stephanie Convery, 'Robot recruiters: can bias be banished from AI hiring?' *Guardian* (26 March 2023), https://www.theguardian.com/technology/2023/mar/27/robot-recruiters-can-bias-be-banished-from-ai-recruitment-hiring-artificial-intelligence; 'U.S. warns of discrimination in using artificial intelligence to screen job candidates', *NPR*

(12 May 2022), https://www.npr.org/2022/05/12/1098601458/artificial-intelligence-job-discrimination-disabilities.

2 De las Casas, *Short History*; Pagden, *Lords of All the World*, Chapter 5; Anthony Pagden, *The Fall of Natural Man* (Cambridge University Press, 1986).

3 For a brilliant attack on this assumption see Mike Savage, *The Return of Inequality* (Harvard University Press, 2021).

4 Salomé Viljeon, 'A Relational Theory of Data Governance', *Yale Law Journal* 131:2 (2021) pp. 370–781.

5 'Perhaps': in fact sociologists of time don't agree on this point (Judy Wajcman, *Pressed for Time: The acceleration of life in digital capitalism* (University of Chicago Press, 2015); Hartmut Rosa, *Social Acceleration: A new theory of modernity* (Columbia University Press, 2015)).

6 Lauren Bridges, 'Infrastructural Obfuscation: unpacking the carceral logics of the Ring surveillance assemblage', *Information Communication & Society* 24:6 (2021) pp. 830–49; Caroline Haskins, 'Amazon's Home Security is Turning Everyone into Cops', *Vice* (7 February 2019), https://www.vice.com/en/article/qvyvzd/amazons-home-security-company-is-turning-everyone-into-cops.

7 Kate Conger, 'Give your nudes to Facebook', *Gizmodo* (5 June 2018), https://gizmodo.com/give-your-nudes-to-facebook-1826545511.

8 'Safety Center', *Google* (n.d.), https://safety.google/intl/en_us/pixel/.

9 Benjamin Grosser, 'What do Metrics Want? How quantification prescribes social interaction on Facebook', *Computational Culture* 4 (2014), http://computationalculture.net/what-do-metrics-want/.

10 'Number of smartphone users in South Africa from 2014 to 2023 (in millions)', *Statista* (2019), https://www.statista.com/statistics/488376/forecast-of-smartphone-users-in-south-africa/.

11 Toussaint Nothias, 'Access Granted: Facebook's Free Basics in Africa', *Media, Culture & Society* 42:3 (2020) pp. 329–48.

12 Nesrine Malik, 'How Facebook took over the internet in Africa – and changed everything', *Guardian* (20 January 2022), https://www.theguardian.com/technology/2022/jan/20/facebook-second-life-the-unstoppable-rise-of-the-tech-company-in-africa.

13 'Accelerate your growth', *Open Network for Digital Commerce* (n.d.), https://ondc.org/.

14 Mark Zuckerberg, 'Building Global Community' (originally posted 16 February 2017) https://www.facebook.com/notes/3707971095882612/ [accessed 23 February 2023].

15 Hao, 'The Facebook whistleblower'; see also the *Wall Street Journal* articles in September–October 2021 based on the documents on which Haugen based her testimony, for example Georgia Wells, Jeff Horwitz and Deepa Seetharaman, 'Facebook Knows Instagram is Toxic for Teen Girls, Company Documents Show', *Wall Street Journal* (13 September 2021), https://www.wsj.com/articles/facebook-knows-instagram-is-toxic-for-teen-girls-company-documents-show-11631620739.

16 Dan Milmo, 'Interview: I never wanted to be a whistleblower, but lives were in danger', *Observer* (21 October 2021), https://www.theguardian.com/technology/2021/oct/24/frances-haugen-i-never-wanted-to-be-a-whistleblower-but-lives-were-in-danger.

17 Quoted, Dan Milmo and Clea Skopeliti, 'Teenage girls, body image, and Instagram's perfect storm', *Guardian* (18 September 2021), https://www.

theguardian.com/technology/2021/sep/18/teenage-girls-body-image-and-instagrams-perfect-storm.

18 Kevin Kelly, *The Inevitable: Understanding the 12 Technological Forces That Will Shape Our Future* (Penguin, 2017).

19 Siebel, *Digital Transformation*, p. 18.

20 'Personal Data: The Emergence of a New Asset Class', *World Economic Forum* (January 2011), http://www3.weforum.org/docs/WEF_ITTC_Personal-DataNewAsset_Report_2011.pdf, p. 7; Soumira Dutta and Beñat Bilbao-Osorio, 'The Global Information Technology Report 2012: Living in a hyper-connected world', *World Economic Forum* (2012), http://www3.weforum.org/docs/Global_IT_Report_2012.pdf, pp. 6, 102.

21 Siebel, *Digital Transformation*, p. 46.

22 Joseph Bradley, Joel Barbier and Doug Handler, 'Embracing the Internet of Everything to Capture Your Share of $14.4 trillion', *Cisco* (2013), https://www.cisco.com/c/dam/en_us/about/ac79/docs/innov/IoE_Economy.pdf.

23 'Number of Internet of Things (IoT) connected devices worldwide from 2019 to 2021, with forecasts from 2022 to 2030', *Statista* (July 2021), https://www.statista.com/statistics/1183457/iot-connected-devices-worldwide/; 'Cisco Annual Internet Report (2018–2023)', *Cisco* (9 March 2020), https://www.cisco.com/c/en/us/solutions/collateral/executive-perspectives/annual-internet-report/white-paper-c11-741490.html.

24 Nathaniel Fick et al., 'Confronting Reality in Cyberspace: Foreign Policy for a Fragmented Internet', *Council on Foreign Relations* (July 2022), https://www.cfr.org/report/confronting-reality-in-cyberspace, p. 3, added emphasis.

25 On Haier, see Kokas *Trafficking Data*, p. 178; for China's latest five-year plan, Rogier Creemers et al., 'Translation: 14th five-year plan for national informatization – Dec. 2021', *DigiChina* (24 January 2022), https://digichina.stanford.edu/work/translation-14th-five-year-plan-for-national-informatization-dec-2021/.

26 Emily West, *Buy Now: How Amazon Branded Convenience and Normalized Monopoly* (MIT Press, 2022); Evan Selinger and Darrin Durant, 'Amazon's Ring: Surveillance as a Slippery Slope Service', *Science as Culture* 31:1 (2021) pp. 92–106.

27 Karl Polanyi, *The Great Transformation* (Beacon Press 2001) p. 60, added emphasis.

28 Asef Bayat, *Life and Politics: How ordinary people change the Middle East* (Stanford University Press, 2013).

29 Arda and Akdemir, 'Activist communication design on social media: The case of online solidarity against forced Islamic lifestyle', *Media, Culture & Society* 43:6 (2021) pp. 1076–94.

30 Quoted Malik, 'How Facebook took over the internet in Africa'.

31 James C. Scott, *Weapons of the Weak: Everyday forms of resistance* (Yale University Press, 1990).

32 Zeynep Tufecki, *Twitter and Teargas: The power and fragility of networked protest* (Yale University Press, 2016).

33 Shakuntala Banaji and Ramnath Bhat, *Social Media and Hate* (Routledge, 2022).

34 Eli Pariser, *The Filter Bubble: What the Internet Is Hiding from You* (Penguin, 2011).

35 Catherine Knight-Steele, *Digital Black Feminism* (New York University Press, 2021); Sarah Florini, *Beyond Hashtags* (New York University Press, 2019);

Sarah Jackson, Moya Bailey and Brooke Foucault Welles, *#Hashtag Activism* (MIT Press, 2020).

36 Jonathan Haidt, 'Why the Past 10 Years of American Life Have Been Uniquely Stupid', *The Atlantic* (11 April 2022), https://www.theatlantic.com/magazine/archive/2022/05/social-media-democracy-trust-babel/629369/; Cohen, *Between Truth and Power*, p. 86.

37 Axel Bruns, 'Filter Bubble', *Internet Policy Review* 8:4 (2019) pp. 1–14; Axel Bruns, *Are Filter Bubbles Real?* (Wiley, 2019); Richard Fletcher, Craig T. Robertson, and Rasmus Kleis Nielsen, 'How Many People Live in Politically Partisan Online News Echo Chambers in Different Countries?', *Journal of Quantitative Description: Digital Media* 1 (2021) pp. 1–56.

38 Shanto Iyengar, Gaurav Sood and Yphtach Lelkes, 'Affect, not Ideology: A Social Identity Perspective on Polarization', *Public Opinion Quarterly* 76:3 (2012) pp. 405–31. On the impacts of social media business models, see Ian Bogost and Alexis Madrigal, 'How Facebook Works for Trump', *The Atlantic* (17 April 2020), https://www.theatlantic.com/technology/archive/2020/04/how-facebooks-ad-technology-helps-trump-win/606403/.

39 Maria Ressa, *How to Stand Up to a Dictator* (Ebury Publishing, 2022) p. 137.

40 See for instance the collection edited by Mark Littler and Benjamin Lee, *Digital Extremisms: Readings in Violence, Radicalisation and Extremism in the Online Space* (Palgrave, 2020).

41 'Rabbit Hole', *New York Times* podcast (n.d.), https://www.nytimes.com/column/rabbit-hole.

42 Zeynep Tufekci, 'YouTube, the Great Radicaliser', *New York Times* (10 March 2018), https://www.niemanlab.org/reading/youtube-the-great-radicalizer/.

43 Julia DeCook and Jennifer Forestal, 'Of Humans, Machines, and Extremism: The Role of Platforms in Facilitating Undemocratic Cognition', *American Behavioral Scientist* (2022) pp. 1–20, https://doi.org/10.1177/00027642221103186; Chun, *Discriminating Data*.

44 Shanto Iyengar and Sean J. Westwood, 'Fear and Loathing Across Party Lines: New Evidence on Group Polarization', *American Journal of Political Science* 59:3 (2014) pp. 690–707.

45 Craig Silverman, Ryan Mac, and Pranav Dixit, '"I Have Blood On My Hands": A Whistleblower Says Facebook Ignored Global Political Manipulation', *Buzzfeed* (14 September 2020), https://www.buzzfeednews.com/article/craigsilverman/facebook-ignore-political-manipulation-whistleblower-memo.

46 Craig Silverman et al., 'How Google's ad business funds disinformation around the world', *ProPublica* (29 October 2022), https://www.propublica.org/article/google-alphabet-ads-fund-disinformation-covid-elections.

47 Cohen, *Between Truth and Power*, Chapter 3.

48 Bryan Harris and Hannah Murphy, 'Brazil's Lawmakers to Vote on "Fake News" Bill Opposed by Tech Groups', *Financial Times* (30 April 2023), https://www.ft.com/content/827326e9-7433-4fb4-9fb5-36a76658d106.

49 Meredith Broussard, *Artificial Unintelligence: How computers misunderstand the world* (MIT Press, 2019).

50 Eric Schmidt, 'This is how AI will change the way science gets done', *MIT Tech Review* (5 July 2023) https://www.technologyreview.com/2023/07/05/1075865/eric-schmidt-ai-will-transform-science/.

51 Emily Bender, Timnit Gebru, Angelina McMillan-Major and Shmargaret Shmitchell, 'On the Dangers of Stochastic Parrots: Can Language Models Be

Too Big?', FaCCT 2021 (3–10 March 2021), available from https://doi. org/10/1145/3442188.3445922, p. 617. For interesting discussion of the deeper philosophical support for the Stochastic Parrots critique, see Alan Blackwell, *Modal Codes: Designing Alternatives to AI* (MIT, forthcoming), Chapter 5.

52 Jim Bisbee, Joshua D. Clinton, Cassy Dorff, Brenton Kenkel and Jennifer Larson, 'Artificially Precise Extremism: How Internet-Trained LLMs Exaggerate Our Differences', *SocArXiv* (2 May 2023), https://osf.io/preprints/socarxiv/5ecfa/.

53 Francis Bacon, *The New Organon and Related Writings*, ed., Fulton H. Anderson (Bobbs-Merrill, 1960), Aphorism 84.

54 Rohan Deb Roy, 'Western science long relied on the knowledge and exploitation of colonized peoples. In many ways, it still does', *Smithsonian Magazine* (9 April 2018), https://www.smithsonianmag.com/science-nature/science-bears-fingerprints-colonialism-180968709/.

55 Claude Alvares, 'Science, colonialism and violence: A luddite view', in Ashish Nandy, *Science, Hegemony and Violence* (Oxford University Press, 1988) p. 85.

56 Susan Bordo, *The Flight to Objectivity: Essays on Cartesianism and Culture* (SUNY Press, 1987); Evelyn Fox Keller, *Reflections on Gender and Science* (Yale University Press, 1985); Sandra Harding, *Whose Science? Whose Knowledge?* (Cornell University Press, 1991).

57 McQuillan, *Resisting AI* (2022) p. 10.

58 Rachel Withers, 'Should Robots Be Conducting Job Interviews? A.I. is playing a new role in hiring', *Slate* (5 October 2020), https://slate.com/technology/2020/10/artificial-intelligence-job-interviews.html.

59 Jeffrey Dastin, 'Amazon scraps secret AI recruiting tool that showed bias against women', *Reuters* (11 October 2018), https://www.reuters.com/article/us-amazon-com-jobs-automation-insight/amazon-scraps-secret-ai-recruiting-tool-that-showed-bias-against-women-idUSKCN1MK08G.

60 McQuillan, *Resisting AI*, pp. 29, 37.

61 Paola Ricaurte, 'Ethics for the majority world: AI and the question of violence at scale', *Media, Culture & Society* 44:4 (2022) pp. 726–45.

62 Cathy O'Neil, *Weapons of Math Destruction* (Crown, 2016).

63 Hannah Murphy and Cristina Criddle, 'Meta's AI-driven advertising system splits marketers', *Financial Times* (27 February 2023), https://www.ft.com/content/fc95a0f7-5e4e-4616-9b17-7b72daee6c60.

64 Turkopticon, 'Beware the Hype: ChaptGPT didn't Replace Human Data Annotators' (4 April 2023), https://news.techworkerscoalition.org/2023/04/04/issue-5/.

65 Hao, 'How the AI industry profits from catastrophe'.

66 Mateus Viana Braz, Paola Tubaro and Antonio Casilli, 'Microwork in Brazil: Who are the wokers behind artificial intelligence?' https://diplab.eu/wp-content/uploads/2023/06/Viana-Braz-Tubaro-Casilli_Microwork-in-Brazil_EN.pdf.

67 Adrienne Williams, Milagros Miceli, and Timnit Gebru, 'The exploited labor behind artificial intelligence', *Noema* (13 October 2022), https://www.noemamag.com/the-exploited-labor-behind-artificial-intelligence/; Gray and Suri, *Ghost Work*.

68 Elizabeth Culliford, 'Rohingya refugees sue Facebook for $150 billion over Myanmar violence', *Reuters* (8 December 2021), https://www.reuters.com/

world/asia-pacific/rohingya-refugees-sue-facebook-150-billion-over-myanmar-violence-2021-12-07/.
69 BSR Staff, 'Human Rights due diligence of Meta's impacts in Israel and Palestine', *Business for Social Responsibility* (22 September 2022), https://www.bsr.org/en/our-insights/report-view/meta-human-rights-israel-palestine.
70 Alex Hern, 'Facebook translates "good morning" into "attack them", leading to arrest', *Guardian* (24 October 2017), https://www.theguardian.com/technology/2017/oct/24/facebook-palestine-israel-translates-good-morning-attack-them-arrest.
71 Noble, *Algorithms of Oppression*.
72 Nathan Benaich and Ian Hogarth, 'State of AI Report', *State of AI* (12 October 2021), https://www.stateof.ai/2021-report-launch.html.
73 Luke Munn, 'The Uselessness of AI Ethics', *AI and Ethics* (2022), https://doi.org/10.1007/s43681-022-00209-w, pp. 1–9.
74 'Montreal Declaration for a responsible development of artificial intelligence', University of Montreal (2018), https://www.montrealdeclaration-responsibleai.com/the-declaration.
75 'EU AI Act, recital (3)', *European Commission* (21 April 2021), https://eur-lex.europa.eu/legal-content/EN/TXT/?uri=celex%3A52021PC0206.
76 Senate Commission of Jurists, 'Estrutura Base do Projeto de Lei', https://legis.senado.leg.br/sdleg-getter/documento/download/3a6b6ff7-f05d-4e72-aa09-a7cb8e28f0b4.
77 For example, https://futureoflife.org/open-letter/pause-giant-ai-experiments/ (28 March 2023), https://www.safe.ai/statement-on-ai-risk (30 May 2023).
78 https://www.dair-institute.org/blog/letter-statement-March2023 (31 March 2023).
79 Quoted in Alex Hern, ' "What Should the Limits Be?" The father of ChatGPT on whether AI will save humanity – or destroy it', *Guardian* (7 June 2023) https://www.theguardian.com/technology/2023/jun/07/what-should-the-limits-be-the-father-of-chatgpt-on-whether-ai-will-save-humanity-or-destroy-it.
80 Sathnam Sanghera, *Empireland: How imperialism has shaped modern Britain* (Penguin, 2021).

Chapter 4 – The New Colonial Class

1 'Digital destroyers: How Big Tech sells war on our communities', *Big Tech Sells War* (n.d.), https://bigtechsellswar.com.
2 C.f. Mizue Aizeki, Geoffrey Boyce, Todd Miller, Joseph Nevins, and Miriam Ticktin, *Smart Borders or a Humane World?*, The Immigrant Defense Project's Surveillance, Tech & Immigration Policing Project, and the Transnational Institute (6 October 2021), https://www.tni.org/en/publication/smart-borders-or-a-humane-world; Visiongain Reports, 'Global Border Security market is projected to grow at a CAGR of 7.0% by 2023: Visiongain Reports Ltd,' *GlobeNewswire* (18 November 2022), https://www.globenewswire.com/news-release/2022/11/18/2558941/0/en/Global-Border-Security-market-is-projected-to-grow-at-a-CAGR-of-7-0-by-2033-Visiongain-Reports-Ltd.html.
3 For another use of the term 'class' in this context, see Jenna Burrell and Marion Fourcade, 'The Society of Algorithms', *Annual Review of Sociology* 47 (2021) pp. 213–37.
4 For more detail, see Couldry and Mejias, *The Costs of Connection*, Chapter 2.

5 Brian Duignan, '5 Fast facts about the East India Company', *Britannica* (n.d.), https://www.britannica.com/story/5-fast-facts-about-the-east-india-company; William Dalrymple, 'The East India Company: the original corporate raiders', *Guardian* (4 March 2015), https://www.theguardian.com/world/2015/mar/04/east-india-company-original-corporate-raiders.

6 For user statistics, see https://www.businessofapps.com/data/uber-statistics/. For employee statistics, see https://stockanalysis.com/stocks/uber/employees/.

7 For the long prehistory of such marketing, see Josh Lauer, *Creditworthy: A history of consumer surveillance and financial identity in America* (Columbia University Press, 2017).

8 Gandy, *Panoptic Sort.*

9 Peter Drucker 1998, cited Davenport, *Big Data @ Work*, p. 21.

10 Christopher Ahlberg (CEO of Recorded Future), quoted ibid.

11 Ondi Timoner (dir.), *We Live in Public* (Interloper Films, 2010).

12 Brandon Fischer of Group M Next consultancy, quoted Turow, *The Aisles Have Eyes*, p. 2, added emphasis.

13 Viktor Mayer-Schönberger and Kenneth Cukier, *Big Data: A revolution that will transform how we live, work and think* (John Murray, 2013) p. 96; Cohen, *Between Truth and Power*, p. 72.

14 Andrew McAfee and Erik Brynjolfsson, 'Big Data: the Management Revolution', *Harvard Business Review* (4 October 2012), https://hbr.org/2012/10/big-data-the-management-revolution.

15 Davenport, *Big Data @ Work*, p. 28.

16 All market capitalisations given as on 16 June 2023. Source (unless otherwise stated): Yahoo, 'GSPC Trending Tickers', *Yahoo Finance* (n.d.), https://uk.finance.yahoo.com/quote/.

17 See for figures as of 30 June 2022, 'Amazon.com Announces Second Quarter Results', *Amazon* (28 July 2022), https://s2.q4cdn.com/299287126/files/doc_financials/2022/q2/Q2-2022-Amazon-Earnings-Release.pdf.

18 Michael Brown, 'Apple sold more smart speakers in Q1 than Google did, but its lead might be short lived', *TechHive* (21 April 2021), https://www.techhive.com/article/579347/apple-sold-more-smart-speakers-than-google.html.

19 Aaron Mok, 'Amazon is offering customers $2 per month for letting the company monitor the traffic on their phones', *Business Insider* (5 December 2022), https://www.businessinsider.com/amazon-offering-users-2-dollars-month-for-track-phone-data-2022-12.

20 https://www.statista.com/chart/amp/18819/worldwide-market-share-of-leading-cloud-infrastructure-service-providers/.

21 See source in note 18.

22 'Search Engine Market Share Worldwide – January 2023', *Statcounter* (n.d.), https://gs.statcounter.com/search-engine-market-share.

23 Quoted Turow, *Voice Catchers*, p. 64.

24 To be precise: 2.989 billion monthly active users, as stated in https://datareportal.com/essential-facebook-stats.

25 'Meta earnings Presentation Q2 2022', *Meta* (n.d.), https://s21.q4cdn.com/399680738/files/doc_financials/2022/q2/Q2-2022_Earnings-Presentation.pdf.

26 Hannah Murphy and Cristina Criddle, 'Meta's AI-driven advertising system splits marketers', *Financial Times* (28 February 2023), https://www.ft.com/content/fc95a0f7-5e4e-4616-9b17-7b72daee6c60.

27 Richard Waters, 'Meta's Existential Risks Mount', *Financial Times* (28 July 2022), https://www.ft.com/content/6ba3fa7b-502c-443d-94e5-1c785bb9 b078.

28 Varun Mishra, 'Apple takes 62% of premium market in Q1 2022', *Counter-Point* (23 June 2022), https://www.counterpointresearch.com/apple-takes-62-premium-market-q1-2022/.

29 Mishra, 'Apple takes 62% of premium market in Q1 2022'.

30 David Curry, 'Apple statistics (2023)', *Business of Apps* (11 January 2023), https://www.businessofapps.com/data/apple-statistics/.

31 Michael Bürgi and Ronan Shields, 'Apple is quietly pushing a TV ad product with media agencies', *DigiDay* (12 October 2022), https://digiday.com/media/apple-is-quietly-pushing-a-tv-ad-product-with-media-agencies/.

32 Brandon Vigliarolo, 'Apple sued for collecting user data despite opt-outs', *The Register* (14 November 2022), https://www.theregister.com/2022/11/14/apple_data_collection_lawsuit/.

33 Patrick McGee and Ryan McMorrow, 'Apple's bargain with Beijing: access to China's factories – and consumers', *Financial Times* (8 November 2022), https://www.ft.com/content/31ab6e36-e683-490a-bf94-b83e031f3169.

34 'Digital Economy Report 2019 – Value Creation and Capture: Implications for Developing Countries', *United Nations Conference on Trade and Development* (4 September 2019), https://unctad.org/system/files/official-document/der2019_en.pdf; Gurumurthy and Chami, *Intelligent Corporation*.

35 Manuel Baigorri, Ben Bartenstein and Dong Cao, 'UAE Spy Chief's Firm Buys into ByteDance at $220 Billion Value', *Bloomberg News* (15 March 2023), https://www.bloomberg.com/news/articles/2023-03-15/uae-spy-chief-s-firm-buys-into-bytedance-at-220-billion-value?leadSource= uverify%20wall.

36 Some, recognising the recent growth of TikTok, relabel the BATX after Tik-Tok's parent company, ByteDance, giving BATJ (ByteDance, Alibaba, Tencent and JD.com) an e-commerce company: Jack Qiu, 'The Return of Billiard Balls? US-China Tech War and China's State-Directed Digital Capitalism', *Javnost – the Public* (2023), https://doi.org/10.1080/13183222.20 23.2200695.

37 Yihan Ma, 'Alibaba's annual segment revenue FY 2017-FY2022', *Statista* (14 June 2022), https://www.statista.com/statistics/708630/revenue-of-alibabacom-segment/.

38 Katharina Buchholz, 'China's most popular digital payment services', *Statista* (8 July 2022), https://www.statista.com/chart/17409/most-popular-digital-payment-services-in-china/.

39 'Baidu Announces Second Quarter 2022 Results', *Baidu* (30 August 2022), https://ir.baidu.com/static-files/6a803761-2f5f-4e23-a573-b24c6a0e9d96.

40 Jonathan Dyble, 'NXP, Baidu Sign New IoT Partnership', *Technology Magazine* (17 May 2020) https://technologymagazine.com/ai-and-machine-learning/nxp-baidu-sign-new-iot-partnership; https://dueros.baidu.com/en/index.html.

41 Buchholz, 'China's most popular digital payment services'.

42 '2022 Second Quarter Results Presentation', *Tencent* (17 August 2022), https://static.www.tencent.com/uploads/2022/08/17/d0d74555d-fe436e57a9a3c05a0ded87b.pdf.

43 Tom Mitchell, 'What the eclipse of Tencent by a liquor company says about Xi's China', *Financial Times* (17 October 2022), https://www.ft.com/content/61ac1bb2-63ed-4655-88b1-5d968f15e357.

44 Mishra, 'Apple takes 62% of premium market in Q1 2022'.
45 Quoted in Chinese Yuan, https://finance.yahoo.com/quote/688036.SS/.
46 Seyram Avle, 'Hardware and Data in the Platform Era: Chinese smartphones in Africa', *Media Culture & Society* 44:8 (2022) pp. 1473–89.
47 Jason Wise, 'How many people use Microsoft in 2023?', *EarthWeb* (23 July 2022), https://earthweb.com/how-many-people-use-microsoft/.
48 https://www.statista.com/chart/amp/18819/worldwide-market-share-of-leading-cloud-infrastructure-service-providers/.
49 Quotations in this and succeeding paras from Satya Nadella, *Hit Refresh: The quest to rediscover Microsoft's soul and imagine a better future for everyone* (Harper Business, 2017) pp. 27, 69, 143, 156.
50 Paul N. Edwards, 'Platforms are Infrastructures on Fire', in Thomas S. Mullaney et al. (eds), *Your Computer Is on Fire* (MIT Press, 2020).
51 Rafe Needleman, 'Who Owns Transit Data?', *CNET* (24 August 2009), https://www.cnet.com/tech/tech-industry/who-owns-transit-data/.
52 Thomas Davenport and DJ Patil, 'Data Scientist: The Sexiest Job of the 21st Century', *Harvard Business Review* (October 2012), https://hbr.org/2012/10/data-scientist-the-sexiest-job-of-the-21st-century; Jeff Hammerbacher, quoted by Robert Gehl, 'Sharing, Knowledge Management and Big Data', *European Journal of Cultural Studies* 18:4–5 (2015) p. 420.
53 Quoted Gehl, 'Sharing, Knowledge Management and Big Data', p. 423, from DJ Patil, Building Data Science Teams (O'Reilly Media, 2011).
54 Gurumurthy and Chami, *Intelligent Corporation*.
55 Mirca Madianou, 'Technocolonialism: Digital innovation and data practices in the humanitarian response to refugee crises', *Social Media + Society* 5:3 (2019) pp. 1–13; Linnet Taylor and Dennis Broeders, 'In the name of development: profit, power and the datafication of the Global South', *Geoforum* (2015) pp. 29–37.
56 Sky Ariella, '25+ Telling Diversity in High Tech Statistics', (7 November 2022), https://www.zippia.com/advice/diversity-in-high-tech-statistics/. For the broader forces at work here, see Noble, *Algorithms of Oppression*; Benjamin, *Race After Technology*; Charlton D. McIlwain, *Black Software: The internet & racial justice, from the AfroNet to Black Lives Matter* (Oxford University Press, 2020).
57 https://www.statista.com/statistics/1256194/representation-of-gender-and-ethnic-minorities-tech/. For the broader forces underlying this, see D'Ignazio and Klein, *Data Feminism*; Sasha Costanza Chock, *Design Justice* (MIT Press, 2020); Ben Little and Alison Winch, *The New Patriarchs of Digital Capitalism* (Routledge, 2021).
58 The account that follows is drawn from Jon Keegan and Alfred Ng, 'Who Is Collecting Data from Your Car?', *The Markup* (27 July 2022), https://the-markup.org/the-breakdown/2022/07/27/who-is-collecting-data-from-your-car.
59 Roger Lanctot of Strategic Analytics, quoted Turow, *Voice Catchers*, p. 135.
60 Sam Hind, Max Kanderske and Fernando van der Vlist, 'Making the Car "Platform Ready": How Big Tech Is Driving the Platformization of Automobility', *Social Media & Society* (2022), https://doi.org/10.1177/205630512 21098697, p. 8.
61 Alberto Cevolini and Elena Esposito, 'From pool to profile: social consequences of algorithmic prediction insurance', *Big Data & Society* 7:2 (2020) pp. 1–11.

62 'The Smart Mobility Data Platform', *Otonomo* (n.d.), https://otonomo.io/.

63 Yannick Perticone, Jean-Christophe Graz and Kunz Rahel, 'Datanalysing the uninsured: the coloniality of inclusive insurance platforms', *Competition & Change*, https://doi.org/10.1177/10245294221125849, pp. 1–21, 14.

64 Keegan and Ng, 'Who Is Collecting Data from Your Car?'; Turow, *Voice Catchers*, p. 136.

65 Joann Muller, 'What Tesla knows about you', *Axios* (13 March 2019), https://www.axios.com/2019/03/13/what-tesla-knows-about-you.

66 Martin Landis, 'Data monetization – Tesla's principles for data-driven success', *USU* (19 August 2021), https://blog.usu.com/en-us/tesla-principles-for-data-driven-success.

67 Torben Iversen and Philipp Rehm, *Big Data and the Welfare State* (Cambridge University Press, 2022).

68 Aitor Jiménez, 'Datafying Islamaphobia in Catalonia', *Medium – Data & Society: Points* (23 November 2022), https://points.datasociety.net/datafying-islamophobia-in-catalonia-b1a87f869cab.

69 'Global Trends 2022', *United Nations Refugee Agency* (2022), https://www.unhcr.org/global-trends.

70 Samuel N. Chambers et al., 'Mortality, Surveillance and the Tertiary "Funnel Effect" on the U.S.-Mexico Border: A Geospatial Modeling of the Geography of Deterrence', *Journal of Borderlands Studies* 36:3 (2019) pp. 443–68; Mizue Aizeki et al., *Smart Borders or A Humane World?* (Transnational Institute, 6 October 2021), https://www.tni.org/en/publication/smart-borders-or-a-humane-world.

71 Weronika Stryzyzynska, 'Iranian authorities plan to use facial recognition to enforce new hijab law', *Guardian,* (5 September 2022), https://www.theguardian.com/global-development/2022/sep/05/iran-government-facial-recognition-technology-hijab-law-crackdown.

72 Max Chafkin, *The Contrarian* (Penguin Books, 2021) p. 266.

73 On PayPal, see Little and Winch, *New Patriarchs*, Chapter 5.

74 Dan Milmo, 'Palantir: Trump-backer's data firm that wants a big NHS deal', *Guardian* (21 June 2022), https://www.theguardian.com/society/2022/jun/21/palantir-concerns-over-data-firm-poised-to-be-operating-system-of-nhs; Ian Johnston, 'NHS data specialists oppose Palantir's bid for £480mn contract', *Financial Times* (16 April 2023), https://www.ft.com/content/2574c49d-2aa4-47e8-9923-35c7aeeb2d3f.

75 Johnston, 'NHS data specialists'.

76 Cristina Alaimo and Jannis Kallinikos, 'Organizations Decentered: Data Objects, Technology and Knowledge', *Organization Science* 33:1 (2022) pp. 19–37.

77 'The War Against Immigrants: Trump's Tech Tools Powered by Palantir,' *Mijente* (August 2019), https://mijente.net/wp-content/uploads/2019/08/Mijente-The-War-Against-Immigrants_-Trumps-Tech-Tools-Powered-by-Palantir_.pdf. See also comment from Georgetown Law in response to Advanced Notice of Proposed Rulemaking on Commercial Surveillance and Data Security at Georgetown Law, 'Comment submitted by the Center on Privacy & Technology at Georgetown Law', *Federal Trade Commission* (21 November 2022), https://www.regulations.gov/comment/FTC-2022-0053-1234.

78 'Privacy advocates demand a European survey of Palantir', *Stichting Onderzdek Marktinformatie* (n.d.), https://somi.nl/en/privacy-advocates-demand-a-european-survey-of-palantir.

79 Palantir, 'Ensuring the resettling and safeguarding of refugees fleeing the war in Ukraine', *Palantir Blog* (20 October 2022), http://blog.palantir.com/ensuring-the-resettling-and-safeguarding-of-refugees-fleeing-the-war-in-ukraine-a5a5fcb306fa. More generally, see Daniel Howden et al., 'Seeing stones: Pandemic reveals Palantir's troubling reach in Europe', *Guardian* (2 April 2021), https://www.theguardian.com/world/2021/apr/02/seeing-stones-pandemic-reveals-palantirs-troubling-reach-in-europe.

80 Andrew Iliadis and Kate Acker, 'The Seer and the Seen: Surveying Palantir's Surveillance Platform', *The Information Society* 38:5 (2022) pp. 334–63. Palantir quotes taken from its SEC filing cited there.

81 David Ignatius, 'Opinion: How the algorithm tipped the balance in Ukraine', *Washington Post* (19 December 2022), https://www.washingtonpost.com/opinions/2022/12/19/palantir-algorithm-data-ukraine-war/.

82 Ignatius, 'Algorithm Tipped the Balance'.

83 William Dalrymple, *The Anarchy: The Relentless Rise of the East India Company* (Bloomsbury, 2019) pp. xxiv–xxv.

84 Mariana Mazzucato, *The Entrepreneurial State* (Anthem Press, 2014); Andrew Keen, *The Internet Is not the Answer* (Atlantic Books, 2015) p. 38; Juan O. Freuler, 'The weaponization of private corporate infrastructure', *Global Media and China* (2022), https://doi.org/10.1177/20594364221139729.

85 Felicia Schwartz and Patrick McGee, 'Pentagon splits $9bn cloud computing contract among tech giants', *Financial Times* (7 December 2022), https://www.ft.com/content/55e5675c-6803-48c0-85d9-2097af69c16e.

86 Turow, *Voice Catchers*, p. 53.

87 Yu Hong, 'Pivot to Internet Plus: Moulding China's Digital Economy for Economic Restructuring', *International Journal of Communication* 11 (2017) pp. 1486–1506; Lianrui Jia and Dwayne Winseck, 'Political Economy of Chinese Internet Companies: Financialization, concentration, and capitalization', 80:1 (2018) pp. 30–59; Kokas, *Trafficking Data*.

88 Sun Yu, 'China's central bank struggles to force tech groups to share data with the state', *Financial Times*, (4 November 2022), https://www.ft.com/content/75409a44-6cfb-43e9-be31-776eb814a919.

89 Kokas, *Trafficking Data*, p. 55.

90 Rogier Creemers, 'Planning Outline for the Construction of a Social Credit System (2014–2020)', *China Copyright and Media* (14 June 2014), https://chinacopyrightandmedia.wordpress.com/2014/06/14/planning-outline-for-the-construction-of-a-social-credit-system-2014-2020/.

91 Malavika Jayaram, 'India's Big Brother Project', *Boston Review* (19 May 2014), https://www.bostonreview.net/articles/malavika-jayaram-india-unique-identification-biometrics/.

92 Bruce Schneier, 'Spy agencies are addicted to corporate data load: Bruce Schneier', *Bloomberg UK* (31 July 2013), https://www.bloomberg.com/view/articles/2013-07-31/the-public-private-surveillance-partnership.

93 AWS, 'The Trusted Cloud for Government', *Amazon Web Services* (n.d.), https://aws.amazon.com/government-education/government/.

94 Paz Peña and Joana Varon, 'Not My AI' (2021), https://notmy.ai/news/case-study-plataforma-tecnologica-de-intervencion-social-argentina-and-brazil/.

95 João Magalhães and Nick Couldry, 'Giving by Taking Away: Big tech, data colonialism, and the reconfiguration of social good', *International Journal of Communication* 15 (2021) pp. 343–62.

96 Silvia Masero, 'Biometric Infrastructures and the Indian Public Welfare Distribution system', *South Asia Multidisciplinary Academic Journal (SAMAJ)* 23 (2020) pp. 1–21.

97 Linnet Taylor and Dennis Broeders, 'In the name of development: profit, power and the datafication of the Global South', *Geoforum* 64 (2015) pp. 229–37, 232.

98 Revati Prasad, 'People as data, data as oil: The digital sovereignty of the Indian state', *Information Communication and Society* 25:6 (2022) pp. 801–15.

99 Nathan Heller, 'Estonia, the Digital Republic', *New Yorker* (11 December 2017).

100 Susan Cahill, 'Surveillance Frontierism', *Helioscope* (10 November 2021).

101 Fabiano Maisonnave, 'Musk brought internet to Brazil's Amazon. Criminals love it', *AP News* (16 March 2023), https://apnews.com/article/amazon-musk-starlink-gold-mining-yanomami-7ab40c14375a9f1bb691a809dcc843b4.

102 Alex Horton, 'Channeling "The Social Network", lawmaker grills Zuckerberg on his notorious beginnings', *Washington Post* (11 April 2018), https://www.washingtonpost.com/news/the-switch/wp/2018/04/11/channeling-the-social-network-lawmaker-grills-zuckerberg-on-his-notorious-beginnings/.

103 Quoted Kashmir Hill, 'The Secretive Company that Might End Privacy as We Know It', *New York Times* (18 January 2020), https://www.nytimes.com/2020/01/18/technology/clearview-privacy-facial-recognition.html.

104 Luke O'Brien, 'The Far-Right Helped Create the World's Most Powerful Facial Recognition Technology', *Huffington Post* (7 April 2020), https://www.huffpost.com/entry/clearview-ai-facial-recognition-alt-right_n_5e7d028bc5b6cb08a92a5c48?6p8.

105 2019 Patent quoted Iliadis and Acker, 'The Seer and the Seen'.

106 Hill, 'The Secretive Company'.

107 'Clearview AI Home', *Clearview AI* (n.d.), https://www.clearview.ai/.

108 Alex Pentland, *Social Physics: How social networks can make us smarter* (Penguin Books, 2015) pp. 6–7.

109 Pentland, *Social Physics*, pp. 8, 220, 14.

110 Company details at https://tracxn.com/d/companies/endor/__CSLVvZ8AutpYKQzc7Xl34NS1to0f2U8-yuLLemaA4wU.

111 Turow, *Voice Catchers*, pp. 93–4.

112 Alex Pentland, Alexander Lipton and Thomas Hardjono, *Building the New Digital Economy: Data as Capital* (MIT Press, 2021).

113 David Olusoga, 'The Ties that Bind Us', *Guardian* (28 March 2023), https://www.theguardian.com/news/ng-interactive/2023/mar/28/slavery-and-the-guardian-the-ties-that-bind-us.

114 Viljeon, 'A Relational Theory of Data Governance'.

115 Srishti Jaswal, 'Would you geotag your home for your government? 50 million Indians just did', *RestofWorld* (15 September 2022), https://restofworld.org/2022/would-you-geotag-your-home-for-your-government-50-million-indians-did/.

116 Quoted ibid.

Chapter 5 – Voices of Defiance

1 John Feng, 'Chinese Delivery Driver Sets Himself on Fire to Protest Unpaid Wages', *Newsweek* (12 January 2021), https://www.newsweek.com/chinese-delivery-driver-sets-himself-fire-protest-unpaid-wages-1560805.

2 Our account relies on Hanke, *La humanidad es una*; Bartolomé de las Casas, *A Short Account of the Destruction of the Indies* (Penguin, 1992); Anthony Pagden, 'Introduction', in Las Casas, *Short Account.*

3 David Graeber and David Wengrow, *The Dawn of Everything* (Penguin, 2022).

4 Leanne Betasamosake Simpson, *As We Have Always Done* (University of Minnesota Press, 2021) p. 15, original emphasis.

5 Simpson, *As We Have Always Done*, p. 75, which draws on Naomi Klein, 'Dancing the World into Being: A conversation with Idle No More's Leanne Simpson', *Yes Magazine* (6 March 2013), https://www.yesmagazine.org/social-justice/2013/03/06/dancing-the-world-into-being-a-conversation-with-idle-no-more-leanne-simpson/.

6 https://idlenomore.ca.

7 Simpson, *As We Have Always Done*, pp. 222–3.

8 Ibid, p. 156.

9 Ibid, pp. 19, 8.

10 Enrique Dussel, *The Invention of the Americas: Eclipse of 'the other' and the myth of Modernity* (Continuum, 1995).

11 Eric Williams, *Capitalism and Slavery* (Penguin, 2023 [o.p.1944]) p. 53, quoting Malachy Postlethwayt, *The African Trade, the Great Pillar and Support of the British Plantation Trade in North America* (J. Robinson, 1745).

12 Williams, *Capitalism and Slavery*, pp. 6–7, 20.

13 Sven Beckert, *Empire of Cotton* (2014) p. xv; More generally, see Sven Beckert and Seth Rockman, *Slavery's Capitalism* (University of Pennsylvania Press, 2016); W.E.B. Du Bois, *Black Reconstruction in America* (Basic Books, 1998).

14 We summarise here Aníbal Quijano, 'Coloniality and Modernity/Rationality', *Cultural Studies* 21:2–3 (2007) p. 177 [originally published in Spanish in 1989].

15 Ramón Grosfoguel, 'Del "Extractivismo Económico" al "Extractivismo Epistémico" y Al "Extractivismo Ontológico": una forma destructiva de conocer, ser y estar en el mundo', *Tabula Rasa* 24 (2016) pp. 123–43. For an excellent critique of this point in English, see Kevin Ochieng Okoth, 'Decolonisation and its Discontents: Rethinking the Cycle of National Liberation', *Salvage* (22 September 2021), https://salvage.zone/decolonisation-and-its-discontents-rethinking-the-cycle-of-national-liberation/.

16 David Scott, 'A Re-enchantment of humanism: and interview with Sylvia Wynter', *Serendip Studio* (2000), https://serendipstudio.org/oneworld/system/files/WynterInterview.pdf; Sylvia Wynter, 'Unsettling the Coloniality of Being/Truth/Power/Freedom: Towards the Human, After Man, Its Overrepresentation – an Argument', *CR: The New Centennial Review* 3:3 (2003) pp. 257–337.

17 Wynter, 'Unsettling the Coloniality', 283ff.

18 Scott, 'A Re-enchantment of humanism', pp. 191, 195.

19 Norbert Wiener, *Cybernetics: or control and communication in the animal and the machine* (Martino Fine Books, 2013 [1948]), pp. 26–7.

20 Stuart Russell, *Human Compatible: AI and the problem of control* (Allen Lane, 2019), pp. 136–8.
21 Russell, *Human Compatible*, pp. 137–9; compare Robin Mansell, *Imagining the Internet: communication, innovation, and governance* (Oxford University Press, 2012).
22 Ian McEwan, *Machines Like Me* (Jonathan Cape, 2019); Rory Cellan-Jones, 'Stephen Hawking warns artificial intelligence could end mankind', *BBC* (2 December 2014), https://www.bbc.com/news/technology-30290540.
23 Wiener, *Cybernetics*, pp. 27–8.
24 Wiener, *Cybernetics*, pp. 28–9.
25 Joseph Weizenbaum, *Computer Power and Human Reason,* (Penguin, 1976) pp. 3–6.
26 Weizenbaum, *Computer Power*, pp. 38, 238.
27 Ibid, p. 265.
28 Ibid, p. 13, original emphasis.
29 Ibid, p. 249.
30 Ibid, pp. 28–9.
31 Achille Mbembe, *On the Postcolony* (University of California Press, 2001).
32 Mbembe, *Critique of Black Reason* (Duke University Press, 2017) p. 1.
33 Mbembe, *Brutalisme* (La Decouverte, 2023) p. 23 (the original French phrase is '*ce devenir-artificiel de l'humanité*').
34 For critique, see David Marriott, 'The becoming-black of the world? On Achille Mbembe's Critique of Black Reason', *Radical Philosophy* 202 (June 2018) pp. 62–71. The larger issues Marriott raises do not, however, affect the broader point about data we develop from Mbembe's work.
35 Mbembe, *Critique of Black Reason*, pp. 5–6, original emphasis.
36 Ibid, p. 179.
37 Ibid, p. 182.
38 Naomi Klein, *The Shock Doctrine: The rise of disaster capitalism* (Penguin, 2008).
39 Naomi Klein, 'Screen New Deal', *The Intercept* (8 May 2020), [n.p.] https://theintercept.com/2020/05/08/andrew-cuomo-eric-schmidt-coronavirus-tech-shock-doctrine/.
40 Naomi Klein, *This Changes Everything* (Penguin, 2014) p. 270, cf 269.
41 Klein, *This Changes Everything*, p. 171.
42 Klein, 'Screen New Deal', [n.p.], quoting Anuja Sonalker, CEO of Steer Tech.
43 Ibid.

Chapter 6 – A Playbook for Resistance

1 Nathan Sheard and Adam Schwartz, 'The movement to ban government use of face recognition', *Electronic Frontier Foundation* (5 May 2022), https://www.eff.org/deeplinks/2022/05/movement-ban-government-use-face-recognition.
2 Avi Asher-Schapiro, 'California City Bans Predictive Policing in U.S. First', *Reuters* (24 June 2020), https://www.reuters.com/article/us-usa-police-tech-trfn-idUSKBN23V2XC; 'Seattle Adopts Nation's Strongest Regulations for Surveillance Technology', *ACLU Washington* (8 August 2017), https://www.aclu-wa.org/news/seattle-adopts-nation%E2%80%99s-strongest-regulations-surveillance-technology.
3 'Reclaim your face', *Reclaim Your Face* (n.d.), https://reclaimyourface.eu/.

4 Paresh Dave, 'US cities are backing off banning facial recognition as crime rises', *Reuters* (12 May 2022), https://www.reuters.com/world/us/us-cities-are-backing-off-banning-facial-recognition-crime-rises-2022-05-12/.

5 Cedric Whitney, Teresa Naval, Elizabeth Quepons, Simrandeep Singh, Steven Rick and Lilly Irani, 'HCI Tactics for Politics from Below: Meeting the Challenges of Smart Cities', Proceeds of CHI 2021 conference (8–13 May 2021), https://doi.org/10.1145/3411764.3445314.

6 David Davidson et al., 'The algorithm addiction,' (20 December 2022), https://www.lighthousereports.nl/investigation/the-algorithm-addiction/.

7 Heather Stewart, '"We're not going away" Strikers bullish about battle for unionisation at Amazon', *Guardian* (12 June 2023) https://www.theguardian.com/technology/2023/jun/11/amazon-coventry-strike-trio-bullish-union; Levi Sumagaysay, 'This obscure band of Facebook workers is in the middle of a heated union fight', *MarketWatch* (26 August 2022), https://www.marketwatch.com/story/this-small-largely-unknown-band-of-facebook-workers-is-in-the-middle-of-a-heated-union-fight-11661542994.

8 Nicole Carpenter, 'The rise of the video game union', *Polygon* (5 December 2022), https://www.polygon.com/23485977/video-game-unions-guide-explainer.

9 Peter Guest, '"We're all fighting the giant": Gig workers around the world are finally organizing', (21 September 2021), https://restofworld.org/2021/gig-workers-around-the-world-are-finally-organizing/.

10 Lauren Kaori Gurley, 'Union membership hit record low in 2022', *Washington Post* (19 January 2023), https://www.washingtonpost.com/business/2023/01/19/union-membership-2022/.

11 Varsha Bansal, 'Gig workers in India are uniting to take back control from algorithms', *Rest of the World* (14 November 2022), https://restofworld.org/2022/gig-workers-in-india-take-back-control-from-algorithms/ and Karen Hao and Nadine Freischlad, 'The gig workers fighting back against the algorithms', *MIT Technology Review* (21 April 2022), https://www.technologyreview.com/2022/04/21/1050381/the-gig-workers-fighting-back-against-the-algorithms/.

12 Anjar Anwar and Mark Graham, 'Hidden Transcripts of the Gig Economy: labour and the new art of resistance among African gig workers', *Environment and Planning A: Economy and Space* 52:7 (2020) pp. 126–1291.

13 Claudia Irizarry Aponte, 'Food Delivery Workers Veer Into Warfare Over', *The City* (5 April 2023), https://www.thecity.nyc/2023/4/5/23670441/app-food-delivery-workers-wage.

14 Udayan Tandon, Vera Khovanskaya, Enrique Arcilla, Mikaiil Haji Hussein, Peter Zschiesche, and Lilly Irani, 'Hostile Ecologies: Navigating the Barriers to Community-Led Innovation', Proc. ACM Hum.-Comput. Interact., p. 6, CSCW2, Article 443 (November 2022), https://escholarship.org/uc/item/6xj932f8.

15 Movimento dos Trabalhadores Sem Teto, *Homeless Worker Movement in Brazil and the struggle for digital sovereignty* (25 April 2023) p. 10, https://nucleodetecnologia.com.br/docs/Cartilha-MTSTec-ENG.pdf

16 Nitasha Tiku, 'Why tech worker dissent is going viral', *Wired* (29 June 2018), https://www.wired.com/story/why-tech-worker-dissent-is-going-viral/.

17 'Automated Apartheid: How facial recognition fragments, segregates and controls Palestinians in the OPT', Amnesty International (2 May 2023), https://www.amnesty.org/en/documents/mde15/6701/2023/en/.

18 'Israeli spyware facilitates human rights violations', *BDS Movement* (n.d.), https://bdsmovement.net/israeli-spyware-facilitates-human-rights-violations#action.

19 BDS Movement, 'Israeli spyware facilitates human rights violations'.

20 Sebastián Lehuedé, 'Big Tech's new headache: data centre activism flourishes across the world,' *LSE Blogs*, (2 November 2022), https://blogs.lse.ac.uk/medialse/2022/11/02/big-techs-new-headache-data-centre-activism-flourishes-across-the-world/.

21 https://www.odbproject.org/.

22 Yásnava E.A. Gil, 'A modest proposal to save the world', *The Rest of the World* (9 December 2020), https://restofworld.org/2020/saving-the-world-through-tequiology/.

23 See for example: Emma Tranter, 'A new streaming platform offers a trove of Inuit content on demand', *Arctic Today* (22 June 2022), https://www.arctic-today.com/a-new-streaming-platform-offers-a-trove-of-inuit-content-on-demand/; Jens Korff, '30 Aboriginal apps you probably didn't know about', *Creative Spirits* (25 April 2023), https://www.creativespirits.info/resources/apps; Kyle Keeler, 'How Wikipedia Erases Indigenous History', *Slate* (2 February 2023), https://slate.com/technology/2023/02/wikipedia-native-american-history-settler-colonialism.html.

24 Peter Austin and Julia Sallabank, 'Introduction', in Peter Austin and Julia Sallabank (eds), *Cambridge Handbook of Endangered Languages* (Cambridge University Press, 2011) p. 2.

25 Karen Hao, 'A new vision of artificial intelligence for the people', *MIT Technology Review* (22 April 2022), https://www.technologyreview.com/2022/04/22/1050394/artificial-intelligence-for-the-people/.

26 Winthrop Rodgers, 'The mission to preserve Sorani Kurdish by getting it onto Google Translate', *Rest of World* (15 May 2023), https://restofworld.org/2023/google-translate-sorani-kurdish-volunteers/.

27 As documented by the Rising Voices project see: 'Digital Security+ Language', *Rising Voices* (n.d.), https://rising.globalvoices.org/digital-security-language/.

28 Cf. Roger Mueller, '76% Ignore Cookie Banners – The User Behavior After 30 Days of GDPR', *Advance Metrics* (20 July 2018) https://www.advance-metrics.com/en/blog/76-ignore-cookie-banners-the-user-behavior-after-30-days-of-gdpr/; and Joe Nocera, 'How cookie banners backfired,' *New York Times*, (29 January 2022), https://www.nytimes.com/2022/01/29/business/dealbook/how-cookie-banners-backfired.html.

29 Soo Youn, ' "I get better sleep": the people who quit social media', *Guardian*, (10 February 2021), https://www.theguardian.com/lifeandstyle/2021/feb/10/people-who-quit-social-media.

30 Felicia Hou, 'Lush CEO says he's "happy to lose" 13 million quitting social media. "We're talking about suicide, not whether someone should dye their hair blonde" ', *Fortune* (1 December 2021), https://fortune.com/2021/12/01/lush-quit-social-media-executive/.

31 Alex Vadukul, ' "Luddite" Teens don't want your likes', *New York Times* (15 December2022),https://www.nytimes.com/2022/12/15/style/teens-social-media.html.

32 Jathan Sadowski, 'I'm a Luddite. You should be one too', *The Conversation* (9 August 2021), https://theconversation.com/im-a-luddite-you-should-be-one-too-163172.

33 As always, Wikipedia is a good place to start learning about decolonisation. See: 'Decolonization', *Wikipedia* (22 February 2023), https://en.wikipedia.org/wiki/Decolonization [accessed 25 February 2023].

34 Jenny Odell, *How to do nothing: resisting the attention economy* (Penguin Books, 2020) p.12.

35 Veronica Gago, *Feminist International: How to change everything* (Verso 2022) pp. 96, 114, 145.

36 Gago, *Feminist International*, pp. 84, 86.

37 Ibid, pp. 10, 48.

38 Ibid, p. 24.

39 'G8 Open Data Charter and Technical Annex,' *Cabinet Office*, (18 June 2013), https://www.gov.uk/government/publications/open-data-charter/g8-open-data-charter-and-technical-annex.

40 'Using local data to address structural racism', *Urban Institute* (n.d.), https://puttinglocaldatatowork.urban.org/using-local-data-address-structural-racism.

41 'Datos contra el feminicídio', *Data Against Feminicide* (August 2019), https://datoscontrafeminicidio.net/.

42 cf Graeber and Wengrow, *The Dawn of Everything*.

43 Ignacio M. Cristi, 'Autogestión Compleja en un Movimiento Urbano Popular: Prefigurando el Buen Vivir en una lucha contra, sin y desde el Estado-Mercado', *RED Pilares 2018* (September 2018), https://www.researchgate.net/publication/327903249_Autogestion_Compleja_en_un_Movimiento_Urbano_Popular_Prefigurando_el_Buen_Vivir_en_una_lucha_contra_sin_y_desde_el_Estado-Mercado.

44 GDPR, recital 1. Available from https://eur-lex.europa.eu/legal-content/EN/TXT/PDF/?uri=CELEX:32016R0679.

45 Hannah Murphy and Javier Espinosa, 'Facebook owner Meta hit with record €1.2 billion fine over EU-US data transfers', *Financial Times* (22 May 2023), https://www.ft.com/content/d1607121-0a2e-4b74-b690-d368d0c290e8; Lisa O'Carroll, 'MEPs to vote on proposed ban of "Big Brother" AI facial recognition on streets', *Guardian* (10 May 2023) https://www.theguardian.com/technology/2023/may/10/meps-to-vote-on-proposed-ban-on-big-brother-ai-facial-recognition-on-streets.

46 Jason Del Rey, 'Amazon's antitrust settlement in Europe sure looks like a win for Amazon', *Vox* (22 December 2022), https://www.vox.com/recode/2022/12/22/23522734/amazon-eu-settlement-buy-box-sellers-antitrust.

47 Mitzi Lázsló, Gregory Narr, Velislava Hillman, Nick Couldry and Russell Newman, '4 Ways the EU Digital Acts Fall Short and How to Remedy It', *Medium* (5 July 2022), https://medium.com/@gregerwinnarr/4-ways-the-new-eu-digital-acts-fall-short-and-how-to-remedy-it-d16b681a88bc; Tamar Sharon and Raphael Gellert, 'Regulating Big Tech expansionism? Sphere transgressions and the limits of Europe's digital regulatory strategy', *Information Communication & Society* (forthcoming).

48 Ian Johnston, 'NHS data specialists oppose Palantir's bid for £480mn contract', *Financial Times* (16 April 2023), https://www.ft.com/content/2574c49d-2aa4-47e8-9923-35c7aeeb2d3f.

49 Emma Llansó, 'Coalition Letter On Privacy and Free Expression Threats in Kids Online Safety Act', *Center for Democracy and Technology* (28 November 2022), https://cdt.org/insights/coalition-letter-on-privacy-and-free-expression-threats-in-kids-online-safety-act/.

50 ACLU, *Submission to Federal Trade Commission on Commercial Surveillance.*
51 Jey Greene and Rachel Lerman, 'Amazon seeks recusal of FTC Chair Khan, a longtime company critic', *Washington Post* (30 June 2021), https://www.washingtonpost.com/technology/2021/06/30/amazon-khan-ftc-recusal/.
52 Ulises Ali Mejias, 'Why the Global South should nationalise its data', *Al Jazeera* (14 December 2019), https://www.aljazeera.com/opinions/2019/12/14/why-the-global-south-should-nationalise-its-data; Leonardo Fabián and Sofía Scasserra, 'La cuestión de la dependencia: estado empresario, planificación constitucional de datos, cosejo económico y social inteligente', *Friederich Ebert Stiftung* (November 2021), https://library.fes.de/pdf-files/bueros/argentinien/18548.pdf.
53 Ben Tarnoff, 'Tech Workers Versus the Pentagon,' *Jacobin* (6 June 2018), https://jacobin.com/2018/06/google-project-maven-military-tech-workers.
54 Nancy Skinner, 'Act Locally: Using City Government for Social Change', in Brad Erickson (ed.), *Call to Action: Handbook for Ecology, Peace and Justice* (Sierra Club Books, 1990) pp. 19–21.
55 Lilly Irani and Khalid Alexander, 'The Oversight Bloc', *Logic* (25 December 2021), https://logicmag.io/beacons/the-oversight-bloc/.
56 Andrea Miller, 'How companies like Amazon, Nike and FedEx avoid paying federal taxes', *CNBC* (14 April 2022), https://www.cnbc.com/2022/04/14/how-companies-like-amazon-nike-and-fedex-avoid-paying-federal-taxes-.html; Karen Weise, 'Amazon's profit soars 220 percent as pandemic drives shopping online', *New York Times* (12 May 2021), https://www.nytimes.com/2021/04/29/technology/amazons-profits-triple.html.
57 Mark Scott and Emily Birnbaum, 'How Washington and Big Tech won the global tax fight', *Politico* (June 30, 2021), https://www.politico.eu/article/washington-big-tech-tax-talks-oecd/. Howard Gleckman, Maryland's Digital Advertising Levy Sets Off a New Battle Over Taxing E-Commerce, *Tax Policy Center* (February 25, 2021), https://www.taxpolicycenter.org/taxvox/marylands-digital-advertising-levy-sets-new-battle-over-taxing-e-commerce.
58 Ressa, *How to Stand Up*, pp. 267–8.
59 Dillon Reisman et al., 'Algorithmic Impact Assessments: A Practical Framework for Public Agency Accountability', *AI Now Institute* (April 2018), https://ainowinstitute.org/aiareport2018.html.
60 https://www.adalovelaceinstitute.org/resource/aia-user-guide/
61 Audre Lorde, 'The Master's Tools Will Never Dismantle the Master's House', in Audre Lorde, *Sister Outsider: Essays and Speeches*, (Crossing Press, 1984).
62 'Chapter Sixty-Two: The colonial hangovers of India's new Telecommunications Bill', *TypeRight – The Digital Nukkad* (3 October 2022), https://typeright.stck.me/post/43411/Chapter-Sixty-Two-The-colonial-hangovers-of-India-s-new-Telecommunications-Bill.
63 Pablo Casanova, 'Internal Colonialism and National Development', *Studies in International Comparative Development* 1:4 (1965) pp. 27–37.
64 We discuss the question of autonomy in terms of the concept of the 'space of the self' in Couldry and Mejias, *Costs of Connection*, Chapter 5.
65 'Solid: your data, your choice', *Solid* (n.d.), https://solidproject.org/ and Jaron Lanier, 'Jaron Lanier Fixes the Internet', *New York Times* (n.d.), https://www.nytimes.com/interactive/2019/09/23/opinion/data-privacy-jaron-lanier.html.

66 Thomas Stackpole, 'What Is Web3?,' *Harvard Business Review* (10 May, 2022), https://hbr.org/2022/05/what-is-web3.
67 Ben Tarnoff, *Internet for the People: The fight for our digital future* (Verso Books, 2022). Compare James Muldoon, *Platform Socialism: How to reclaim our digital future from Big Tech* (Pluto Press, 2021).
68 Scholz and Schneider, *Ours to Hack and to Own.*
69 Jeremy Gilbert, *Twenty-first Century Socialism* (Verso 2020); James Muldoon, *Platform Socialism: How to Reclaim our Digital Future from Big Tech* (Pluto 2022); Jeremy Gilbert and Alex Williams, *Hegemony Now: How Big Tech and Wall Street Won the World (And How We Win It Back)* (Verso 2022).
70 These strategies are adapted from Richard Grossman, 'Introduction: Creating Cultures of Resistance', in Brad Erickson (ed.), *Call to Action: Handbook for Ecology, Peace and Justice* (Sierra Club Books, 1990) pp. 8–11.
71 'Algorithm Watch', *AlgorithmWatch* (n.d.), https://algorithmwatch.org/en/ and 'The Markup', *The Markup* (n.d.), https://themarkup.org/.
72 Turow, *Aisles Have Eyes.*
73 https://en.wikipedia.org/wiki/Unionization_in_the_tech_sector (last modified 4 July 2023).
74 Karen Hao et al., 'AI Colonialism', article series in *MIT Technology Review* (April 2022), https://www.technologyreview.com/supertopic/ai-colonialism-supertopic; Rigoberto Lara Guzmán, Ranjit Singh and Patrick Davison, 'Parables of AI in/from the Majority World', *Data & Society* (7 December 2022), https://datasociety.net/library/parables-of-ai-in-from-the-majority-world-an-anthology/.
75 'Código doméstico in the flesh', *Código Doméstico* (n.d.), http://www.codigo domestico.com.
76 'Educação Vigiada', *Educação Vigiada* (n.d.), https://educacaovigiada.org.br.
77 'UK police fail to meet "legal and ethical standards" in use of facial recognition', *University of Cambridge* (n.d.), https://www.cam.ac.uk/research/news/uk-police-fail-to-meet-legal-and-ethical-standards-in-use-of-facial-recognition.
78 https://datajusticelab.org/.
79 Umberto Eco, *Apocalypse Postponed* (Indiana University Press, 1994).
80 Ulises A. Mejias, *Off the Network: Disrupting the digital world* (University of Minnesota Press, 2013).
81 Paola Ricaurte and Rafael Grohmann, 'Data sovereignty and alternative development models', *Bot Populi* (22 October 2021), https://botpopuli.net/data-sovereignty-and-alternative-development-models/.
82 'GeoComunes', *GeoComunes* (n.d.), http://geocomunes.org/; 'InfoAmazônia', *InfoAmazonia* (n.d.), https://infoamazonia.org; 'MariaLab', *MariaLab* (n.d.), https://www.marialab.org/.
83 Paulo Freire, *Pedagogy of the Oppressed*, (Penguin, 2017).
84 Antonio Gramsci, *Selections from Political Writings, 1910–1920. Vol. 1.* Trans. John Mathews, ed. Quintin Hoare, (Lawrence and Wishart, 1977).
85 Sara Ahmed, *Living a Feminist Life* (Duke University Press, 2017) p. 252.
86 Alison Powell, *Undoing Optimization: Civic action in smart cities* (Yale University Press, 2021).
87 But see https://www.tierracomun.net/integrantes.

NOTES

Conclusion – And If We Don't Resist?

1 Emily Tucker, 'Our Future Inside The Fifth Column – Or, What Chatbots Are Really For', *Tech Policy Press* (14 June 2023), https://techpolicy.press/our-future-inside-the-fifth-column-or-what-chatbots-are-really-for/.
2 Mira Murati, quoted in Madhumita Murgia, 'Open AI's Murati: the woman charged with pushing generative AI into the real world', *Financial Times* (18 June 2023), https://www.ft.com/content/73f9686e-12cd-47bc-aa6e-52054708b3b3.
3 Ryan Calo and Danielle K. Citron, 'The Automated Administrative State: A Crisis of Legitimacy', *Emory Law Journal* 70:4 (2020) pp. 798–845.
4 Quoted in Chan Ho-him, 'AI-enabled teddies could tell children bedtime stories, says toymaker', *Financial Times* (16 June 2023), https://www.ft.com/content/acf0307c-ca6d-445d-889a-50cbe64d61e2. The so-called 'Internet of Toys' has been in development already for some years: Donell Holloway and Leila Green (2016) 'The Internet of Toys', *Communication Research and Practice* (2:4) pp. 506–19.
5 We don't have time to go into the academic debate on the term 'autonomy', but see Couldry and Mejias, *Costs*, Chapter 5; Beate Rössler, *Autonomy* (Polity, 2021).
6 Allen W. Wood, *Hegel's Ethical Thought* (Cambridge University Press, 1990) p. 51.
7 Nita Farahany, *The Battle for Your Brain: Defending the Right to Think Freely in an Age of Neurotechnology* (St. Martin's Press, 2023).

FURTHER READING
SUGGESTIONS

Our book draws on a vast literature of academic and policy debates. Each chapter's detailed references will give you many specific leads, but the suggestions below bring together some accessible key resources, organised by theme, that could be general starting points for your own further reading.

Items are listed by alphabetical order, author's last name.

Artificial Intelligence

Meredith Broussard, *Artificial Unintelligence: How Computers Misunderstand the World* (MIT Press, 2018).

Rick Claypool and Cheyenne Hunt, ' "Sorry in Advance!" Rapid Rush to Deploy Generative AI Risks a Wide Array of Automated Harms', (18 April 2023), https://www.citizen.org/article/sorry-in-advance-generative-ai-artificial-intellligence-chatgpt-report/.

Amba Kak and Sarah Myers West, 'AI Now 2023 Landscape: Confronting Tech Power', AI Now Institute (11 April 2023), https://ainowinstitute.org/2023-landscape.

Paola Ricaurte, 'Ethics for the Majority World: AI and the Question of Violence at Scale', *Media, Culture & Society* 44:4 (1 May 2022) pp. 726–45, https://doi.org/10.1177/01634437221099612.

Stuart Russell, *Human Compatible: Artificial Intelligence and the Problem of Control* (Viking, 2019).

Joseph Weizenbaum, *Computer Power and Human Reason* (Harmondsworth, 1976).

Norbert Wiener, *Cybernetics* (Martino Publishing, 2013) [originally published in 1948].

Adrienne Williams, Milagros Miceli and Temnit Gebru, 'The Exploited Labor Behind Artificial Intelligence', *Noema* (13 October 2022), https://www.noemamag.com/the-exploited-labor-behind-artificial-intelligence/.

Colonialism

In addition to the classic texts on colonialism (for example, Chinua Achebe, Bartolome de las Casas, Aimé Césaire, Frantz Fanon, Eduardo Galeano, Albert Memmi, Edward Said, Ngugi Wa Thiong'o), here are some more recent starting points:

Gurminder Bhambra and John Holmwood, *Colonialism and Modern Social Theory* (Routledge, 2022).

Naomi Klein, 'Dancing the World into Being: A Conversation with Idle No More's Leanne Simpson', *YES! Magazine* (5 March 2013), http://www.yesmagazine.org/peace-justice/dancing-the-world-into-being-a-conversation-with-idle-no-more-leanne-simpson.

Achille Mbembe, *Critique of Black Reason* (Duke University Press, 2017).

Sathnam Sanghera, *Empireland: How Imperialism Has Shaped Modern Britain* (Penguin, 2021).

Sofia Scasserra and Carolina Martínez Elebi, 'Digital Colonialism: Analysis of Europe's Trade Agenda', *Transnational Institute* (6 October 2021), https://www.tni.org/en/publication/digital-colonialism.

Leanne Betasamosake Simpson, *As We Have Always Done* (University of Minnesota Press, 2017).

Data Extraction

Mark Andrejevic, *Infoglut: How Too Much Information Is Changing the Way We Think and Know* (Routledge, 2013).

Nick Couldry and Ulises A. Mejias, *The Costs of Connection: How Data Is Colonizing Human Life and Appropriating it For Capitalism* (Stanford University Press, 2019).

Christian Katzenbach and Thomas Christian Bächle, 'Defining Concepts of the Digital Society,' *Internet Policy Review* 8:4 (30 November 2019), https://policyreview.info/concepts/defining-concepts-digital-society [special journal issue that is open access].

Andrew Keen, *The Internet Is not the Answer* (Atlantic Books, 2016).

Aynne Kokas, *Trafficking Data: How China is Winning the Battle for Digital Sovereignty* (Oxford University Press, 2022).

Sarah Lamdan, *Data Cartels: The Companies That Control and Monopolize Our Information* (Stanford University Press, 2022).

Evgeny Morozov, *The Net Delusion: How Not to Liberate the World* (Allen Lane, 2011).

Nick Srnicek, *Platform Capitalism* (Polity, 2016).

Zoetanya Sujon, *The Social Media Age* (Sage, 2021).

Jose Van Dijck, Thomas Poell and Martijn de Wajl, *Platform Societies* (Oxford University Press, 2019).

Shoshana Zuboff, *The Age of Surveillance Capitalism: The Fight for a Human Future at the New Frontier of Power* (Profile Books, 2019).

Data Extraction in Particular Sectors

Mizue Aizeki, Geoffrey Boyce, Todd Miller, Joseph Nevins, & Miriam Ticktin, 'Smart Borders or a Humane World?' *The Immigrant Defense Project's Surveillance, Tech & Immigration Policing Project, and the Transnational Institute* (October 2021), https://www.tni.org/en/publication/smart-borders-or-a-humane-world.

Sarah Brayne, *Predict and Surveil: Data, Discretion and the Future of Policing* (Oxford University Press, 2020).

Kelly Bronson, *The Immaculate Conception of Data: Agribusiness, Activists and Their Shared Vision of the Future* (McGill-Queen's University Press, 2022).

Kelly Gates, *Our Biometric Future: Facial Recognition Technology and the Culture of Surveillance* (Oxford University Press, 2011).

Ilona Kickbusch, Anurag Agrawal, Andrew Jack, Naomi Lee, and Richard Horton, 'Governing health futures 2030: Growing up in a digital world – a joint The Lancet and Financial Times Commission,' *The Lancet* 394, no. 10206 (20 September 2019), https://www.thelancet.com/journals/lancet/article/PIIS0140-6736 (19)32181-6/fulltext.

Joseph Turow, *The Voice Catchers* (Yale University Press, 2021) [on voice profiling and marketing].

Ben Williamson, *Big Data in Education* (Sage, 2017).

Democracy and Human Rights

Lina Dencik, Arne Hintz, Joanna Redden, and Emiliano Treré, *Data Justice*, (Sage, 2022).

Juan Ortiz Freuler, 'Datafication and the Future of Human Rights Practice', *Just Labs* (2021), https://www.openglobalrights.org/datafication-report/.

Tarleton Gillespie, *Custodians of the Internet: Platforms, Content Moderation, and the Hidden Decisions That Shape Social Media* (Yale University Press, 2018).

Maria Ressa, *How to Stand Up to a Dictator: The Fight for Our Future* (W.H. Allen, 2022).

Sarah Roberts, *Behind the Screen: Content Moderation in the Shadows of Social Media* (Yale University Press, 2019).

Siva Vaidhyanathan, *Antisocial Media: How Facebook Disconnects Us and Undermines Democracy* (Oxford University Press, 2018).

Economics

Sven Beckert and Seth Rockman (eds), *Slavery's Capitalism: A New History of American Economic Development* (University of Pennsylvania Press, 2016).

Matthew Desmond, 'In order to understand the brutality of American capitalism, you have to start with the plantation', *New York Times* (18 August 2019), https://www.nytimes.com/interactive/2019/08/14/magazine/slavery-capitalism.html.

Jeremy Gilbert and Alex Williams, *Hegemony Now: How Big Tech and Wall Street Won the World (And How We Win It Back)* (Verso, 2022).

Anita Gurumurthy and Nandina Chami, 'The Intelligent Corporation', *Transnational Institute* (16 January 2020), https://longreads.tni.org/stateofpower/the-intelligent-corporation-data-and-the-digital-economy.

Sandra Mezzadra and Brett Neilson, *The Politics of Operations: Excavating Contemporary Capitalism* (Duke University Press, 2019).

Jason Moore, *Capitalism and the Web of Life* (Verso, 2016).

Eric Williams, *Capitalism and Slavery* (Penguin, 2022) [originally published in 1944].

Environment

Benedetta Brevini, *Is AI Good for the Planet?* (Polity, 2021).

Mel Hogan, 'Big Data Ecologies', *Ephemera* 18:3 (August 2018), https://ephemera journal.org/sites/default/files/2022-01/18-3hogan.pdf.

Naomi Klein, *This Changes Everything: Capitalism vs. The Climate* (Simon & Schuster, 2015).

Richard Maxwell and Toby Miller, *Greening the Media* (Oxford University Press, 2012).

Petr Spelda and Vit Stritecky, 'The Future of Human-Artificial Intelligence Nexus and its Environmental Costs', *Futures* 117 (1 March 2020), https://doi.org/10.1016/j.futures.2020.102531.

Julia Velkova, 'Thermopolitics of data: cloud infrastructures and energy futures', *Cultural Studies* 35:4–5 (2021), https://www.tandfonline.com/doi/full/10.1080/09502386.2021.1895243.

Feminism

Catherine D'Ignazio and Lauren F. Klein, *Data Feminism* (MIT Press, 2020).

Veronica Gago, *Feminist International: How to Change Everything* (Verso, 2020).

Ben Little and Alison Winch, *The New Patriarchs of Digital Capitalism* (Routledge, 2021).

Paola Ricaurte, 'Data Epistemologies, The Coloniality of Power, and Resistance.' *Television & New Media*, 20(4), 350–365 (2019). https://doi.org/10.1177/1527476419831640

Julia Ticona, *Left to Our Own Devices: Coping with Insecure Work in a Digital Age* (Oxford University Press, 2022).

Joana Varon and Paz Peña. *Not My A.I.: Towards Critical Feminist Frameworks to Resist Oppressive A.I. Systems* (Carr Center for Human Rights Policy, Harvard Kennedy School, 2022).

Inequality

Payal Arora, *The Next Billion Users: Digital Life Beyond the West* (Harvard University Press, 2019).

Virginia Eubanks, *Automating Inequality: How High-Tech Tools Profile, Police, and Punish the Poor* (St. Martin's Press, 2018).

Oscar Gandy, *The Panoptic Sort: A Political Economy of Personal Information* (Oxford University Press, 2021) [originally published in 1993].

Mary Madden, Michele Gilman, Karen Levy and Alice Marwick, 'Privacy, Poverty and Big Data: A Matrix of Vulnerabilities for Poor Americans', *Washington University Law Review* 95:1 (2017) pp. 53–125.

Cathy O'Neil, *Weapons of Math Destruction: How Big Data Increases Inequality and Threatens Democracy* (Crown, 2017).

Law, Governance and Regulation

Ryan Calo and Danielle Citron, 'The Automated Administrative State: A Crisis of Legitimacy', *Emory Law Journal* 70:4, pp. 798–845.

Julie Cohen, *Between Truth and Power: The Legal Constructions of Informational Capitalism* (Oxford University Press, 2019).

Louise Hooper, Sonia Livingstone, and Krakae Pothong, 'Problems with Data Governance in UK schools: the cases of Google Classroom and ClassDojo', *Digital Futures Commission and 5Rights Foundation* (August 2022), https://digitalfutures commission.org.uk/wp-content/uploads/2022/08/Problems-with-data-governance-in-UK-schools.pdf.

Stephanie Russo Carroll, Desi Rodriguez-Lonebear, and Andrew Martinez, 'Indigenous Data Governance: Strategies from United States Native Nations', *Data Science Journal* 18:1 (8 July 2019), https://doi.org/10.5334/dsj-2019-031.

Bruce Schneier, *Click Here to Kill Everybody: Security and Survival in a Hyper-Connected World* (W.W. Norton & Company, 2019).

Race

Ruha Benjamin, *Race After Technology: Abolitionist Tools for the New Jim Code* (Polity, 2019).

Wendy Chun, *Discriminating Data: Correlation, Neighborhoods, and the New Politics of Recognition* (MIT 2021).

Jane Chung, 'Racism In, Racism Out: A Primer on Algorithmic Racism,' *Public Citizen* (August 2021), https://www.citizen.org/article/algorithmic-racism/.

Seeta Peña Gangadharan, 'Digital Inclusion and Data Profiling', *First Monday* 17:5 (7 May 2012), https://journals.uic.edu/ojs/index.php/fm/article/view/3821/3199.

Catherine Knight Steele, *Digital Black Feminism* (New York University Press, 2021).

Safiya Umoja Noble, *Algorithms of Oppression: How Search Engines Reinforce Racism* (New York University Press, 2018).

Resistance

Andre Brock, *Distributed Blackness: African American Cybercultures* (New York University Press, 2020).

Sasha Costanza Chock, *Design Justice: Community-Led Practice to Build the Worlds We Need* (MIT Press, 2020).

Ivan Illich, *Tools for Conviviality* (Marion Boyars Publishers, 2021) [originally published in 1972].

Geert Lovink, *Sad by Design: On Platform Nihilism* (Pluto Press, 2019).

Charlton McIlwain, *Black Software* (Oxford University Press, 2020).

Katherine McKittrick (ed.), *Sylvia Wynter: On Being Human as Praxis* (Duke University Press, 2015).

Alison B. Powell, *Undoing Optimization: Civic Action in Smart Cities* (Yale University Press, 2021).

Beate Roessler, *Autonomy: An Essay on the Life Well Lived* (Polity, 2021).

Trebor Scholz and Nathan Schneider, *Ours to Hack and to Own: The Rise of Platform Cooperativism, A New Vision for the Future of Work and a Fairer Internet* (OR Books, 2017).

Ben Tarnoff, *Internet for the People: The Fight for Our Digital Future* (Verso, 2022).

Good sources for finding out about campaigns for resistance include:

https://consentofthenetworked.com/get-involved

https://edri.org/about-us/our-network/

https://en.wikipedia.org/wiki/Category:Digital_rights

https://rankingdigitalrights.org/get-involved/

Work

Ifeoma Ajunwa, *The Quantified Worker: Law and Technology in the Modern Workplace* (Cambridge University Press, 2023).

Mary Gray and Siddharth Suri, *Ghostwork: How to Stop Silicon Valley from Building a New Global Underclass* (Harper Business, 2019).

Karen Hao, 'AI Colonialism', *MIT Technology Review* (20 April 2022), https://www.technologyreview.com/supertopic/ai-colonialism-supertopic.

Karen Levy, *Data Driven: Truck Drivers, Technology and the New Workplace Surveillance* (Princeton University Press, 2023).

Milagros Miceli and Julian Posada, 'The Data-Production Dispositif', *Proc. ACM Hum.-Comput. Interact.* 6, CSCW2, Article 460 (November 2022), https://doi.org/10.1145/3555561.

Trebor Scholz, *Uberworked and Underpaid: How Workers Are Disrupting the Digital Economy* (Polity, 2016).

INDEX

privacy regulations 50, 224
Securities and Exchange Commission
(SEC) 163
Silicon Valley *see* Silicon Valley
slavery and 7, 30, 31, 187
Supreme Court 25–6, 224
taxation in 227–8
undersea telegraph cables and 3, 30
unions in 101, 210, 211
whistle-blowers in 43, 71, 112,
146, 212
workforce layoffs in 51
workplace surveillance in 91, 93
See also individual company name

vaccines 8, 123, 203
Varian, Hal 64–8, 70, 72, 82
Venezuela 130
venture capital 51, 232
Verenigde Oostindische Compagnie
5, 17
Vía Libre 237
violence
AI and 128, 130, 131
borders and 160
colonialism and 6, 29, 34–5, 38,
39–40, 43–5, 102, 183, 184,
188, 198–9, 200, 208
data colonialism and 39–40, 43–5, 144
data violence 43–5
gender-based 204–6
genocidal 8, 131, 179
physical 6, 34–5, 144
symbolic 39–40, 144
violent content 104, 130
VitalityLife 89
voice intelligence industry 165
voice picking 93
voices of defiance 22, 177–204
AI, early apprehensions about 195–8
Bartolomé de Las Casas 179–80
capitalism and colonialism 184–92
coloniality 187–90
delivery drivers 177–9
earlier computer age, warnings from
an 192–8
indigenous critique tradition 180–84
resisting data colonialism in practice
198–203
resources for resistance 203–4
VTech Holdings 245

wages 10, 51, 94, 96, 129–30, 177,
178, 212
water use 26, 77
WeChat 3, 74, 90, 108, 111, 146, 148,
189, 219
Weizenbaum, Joseph 195–7, 204
welfare systems 39, 166–8, 173, 190,
210, 228, 235
Wengrow, David: *The Dawn of
Everything* 180–81
WhatsApp 3, 8, 92, 105, 108, 110,
111, 138, 146, 219, 230.
See also Facebook
Whittaker, Meredith 212
Wiener, Norbert: *Cybernetics* 192–5,
197, 204
Wikipedia 28–9, 214, 234
Williams, Eric: *Capitalism and Slavery*
185–8, 190
Wired 113
women/girls
abortion and 25–7
data discrimination and 25–7, 42–4,
127, 130, 131–2, 153, 167,
169–70, 207, 235
resistance and 218, 235
social media and 43, 112, 116
Wong, Allan 245
work, transformation of 10, 90–97,
173, 206
World Bank 31, 83, 153
World Economic Forum (WEF)
113–14
World Wide Web 63, 230
Wynter, Sylvia 190–92, 197,
200, 204

Xandr 15, 149
Xiaomi 49, 147–9

Yeung, Karen 72
YouTube 3, 120, 169, 170

Zapatistas 221
Zhang, Sophie 71, 121
Zomato 211
zones of extraction 79, 218
Zuboff, Shoshana 32–3, 72
Zuckerberg, Mark 86, 103, 111–13,
169–70, 247
Zucman, Gabriel 31